Multivariate Analyses of Codon Usage Biases

This book is dedicated to Tami and Noboru Sueoka

Statistics for Bioinformatics Set

coordinated by
Guy Perrière

Multivariate Analyses of Codon Usage Biases

Jean R. Lobry

First published 2018 in Great Britain and the United States by ISTE Press Ltd and Elsevier Ltd

ISTE Press Ltd
27-37 St George's Road
London SW19 4EU
UK

www.iste.co.uk

Elsevier Ltd
The Boulevard, Langford Lane
Kidlington, Oxford, OX5 1GB
UK

www.elsevier.com

Notices

Knowledge and best practice in this field are constantly changing. As new research and experience broaden our understanding, changes in research methods, professional practices, or medical treatment may become necessary.

Practitioners and researchers must always rely on their own experience and knowledge in evaluating and using any information, methods, compounds, or experiments described herein. In using such information or methods they should be mindful of their own safety and the safety of others, including parties for whom they have a professional responsibility.

To the fullest extent of the law, neither the Publisher nor the authors, contributors, or editors, assume any liability for any injury and/or damage to persons or property as a matter of products liability, negligence or otherwise, or from any use or operation of any methods, products, instructions, or ideas contained in the material herein.

For information on all our publications visit our website at http://store.elsevier.com/

British Library Cataloguing-in-Publication Data
A CIP record for this book is available from the British Library
Library of Congress Cataloging in Publication Data
A catalog record for this book is available from the Library of Congress
ISBN 978-1-78548-296-0

Printed and bound in the UK and US

Contents

Acknowledgments   ix

Introduction   xi

   I.1. Prerequisites and notations   xi
   I.2. The case under study is a genomic monster   xvii

**Chapter 1. Introduction to Correspondence
Analysis**   1

   1.1. Chapter objectives   1
   1.2. Metric choice   2
      1.2.1. A small data set example   2
      1.2.2. Euclidean distance   7
      1.2.3. Euclidean distance on row profiles   8
      1.2.4. Euclidean distance on double profiles   9
      1.2.5. The χ^2 distance   10
   1.3. Properties   12
      1.3.1. Scree plot of eigenvalues   12
      1.3.2. Factor orientations are meaningless   15
      1.3.3. CA is symmetric   17
      1.3.4. Nonlinearity in CA is not a bug but a feature   17
      1.3.5. Distributional equivalence principle   20
      1.3.6. So, why the χ^2 metric?   20

Chapter 2. Global Correspondence Analysis   25

2.1. Data set   25
    2.1.1. Queries in database   25
    2.1.2. Data polishing   31
2.2. Running global correspondence analysis   34
2.3. The missing factor F_0   38
2.4. First factor   40
    2.4.1. Coding sequence point of view   40
    2.4.2. Codon point of view   42
    2.4.3. Biological interpretation   44
2.5. Second and third factors   44
    2.5.1. Coding sequence point of view   45
    2.5.2. Codon point of view   48
    2.5.3. Biological interpretation   51
2.6. Fourth and fifth factors   53
    2.6.1. Coding sequence point of view   55
    2.6.2. Codon point of view   55

**Chapter 3. Within and Between
Correspondence Analysis**   59

3.1. Running the analyses   59
3.2. Synonymous codon usage (WCA)   63
    3.2.1. The first and unique factor F_1   63
        3.2.1.1. Coding sequences point of view   63
        3.2.1.2. Codon point of view   63
        3.2.1.3. Biological interpretation   65
3.3. Amino acid usage (BCA)   66
    3.3.1. The missing factor F_0   66
    3.3.2. The ugly (F_1, F_2, F_3) *ménage à trois*   66
        3.3.2.1. Coding sequences point of view   66
        3.3.2.2. Aminoacid point of view   69
        3.3.2.3. Methodological point of view   69

Chapter 4. Internal Correspondence Analysis   71

4.1. Running the analyses   71
4.2. Synonymous codon usage   75

4.3. Non-synonymous codon usage   75
    4.3.1. Between-group analysis   75
    4.3.2. Within-group analysis   80
        4.3.2.1. Regular factors   80
        4.3.2.2. Greedy factors   82

Conclusion   87

Appendices   89

Appendix 1   91
  A1.1. Introduction   91
  A1.2. Chapter 1   96
  A1.3. Chapter 2   105
  A1.4. Chapter 3   116
  A1.5. Chapter 4   121

Appendix 2   131
  A2.1. Session information   131

References   133

Index   147

Acknowledgments

Thanks are due to Anne-Béatrice DUFOUR, Laurent GUÉGUEN, Paweł MACKIEWICZ, Simon PENEL, Guy PERRIÈRE and Haruo SUZUKI.

This work was performed using the computing facilities of the CC LBBE/PRABI.

Introduction

I.1. Prerequisites and notations

A basic knowledge of the ℜ statistical software [RCO 13] is assumed[1]. We should be able to install and load the seqinr [CHA 07] and ade4 [CHE 04] packages. In the following, outputs are presented in black to distinguish them from inputs, and comments start with the # character:

```
library(seqinr)
library(ade4)
pi <- 3      # God's approximation in 1 Kings 7:23 and 2 Chronicles 4:2
pi
```

```
[1] 3
```

```
library(fortunes)
fortune("pi <- 3")
```

---

1 There are countless freely available documents for ℜ on the net, including elementary introductions in Chinese, Croatian, Czech, Danish, Farsi, French, German, Greek, Hungarian, Italian, Japanese, Polish, Portuguese, Romanian, Russian, Slovak, Spanish, Ukrainian, Vietnamese and of course *bad English*, which is the language of science according to Jan DE LEEUW in fortune(200).

```
John Fox: I've never understood why it's legal to change the built-in
global "constants" in R, including T and F. That just seems to me to set
a trap for users.
Why not treat these as reserved symbols, like TRUE, Inf, etc.?
Rolf Turner: I rather enjoy being able to set pi <- 3.
    -- John Fox and Rolf Turner
        R-help (June 2013)
```

To avoid too much verbosity, the code used to produce figures is detailed in the Appendix. This book is composed with LATEX using the command Sweave() [LEI 02] that allows for the automatic insertion of ®️ outputs. For instance, the approximation of π that was used by the Babylonians defined in the above code could be referred to in the source text as \Sexpr{pi} and then automatically transformed into three in the final document.

Some code fragments that need an Internet connection or that are time consuming are encapsulated using a logical flag that must be set to TRUE to allow for their execution. The list of these flags is as follows:

```
# NIC: Need Internet Connection
# CTI: Computer Time Intensive
TODO <- FALSE         # Global switch
final.query0 <- TODO     # NIC: get Bb chromosome for chirochore
        structure
final.stability <- TODO  # NIC+CTI: codon usage in 12,317 bacteria
final.query1 <- TODO     # NIC: get sequences and infos from remote
        database
final.query2 <- TODO     # NIC: get ribosomal sequences informations
final.query3 <- TODO     # NIC: get PheL phenylalanin operon leader
        peptide
final.query4 <- TODO     # NIC: get GC content for Ixodes scapulatis CDS
final.screeplot2 <- TODO # CTI: many CA on simulated tables under H0
final.delrow <- TODO     # CTI: many CA while deleting progressively
        small CDS
final.delcol <- TODO     # CTI: many CA while deleting prgressively
        minor codons
final.ttuco.eig <- TODO  # CTI: many WCA and BCA on simulated tables
        under H0
final.ica.eig1 <- TODO   # CTI: many WVA and BCA for within-between
        group analysis
final.ica.eig2 <- TODO   # CTI: many ICA on simulated tables under H0
```

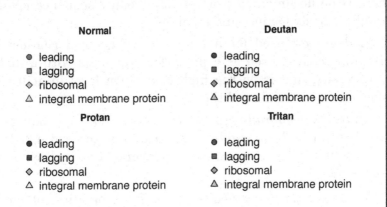

Figure I.1. *Color code. For a color version of this figure, see www.iste.co.uk/lobry/multivariate.zip*

NOTES ON FIGURE I.1.– The colors and the shape of points used in this book to represent different classes of coding sequences are given in the top panel. Simulations with package dichromat [LUM 13] of how these colors are perceived by three common color blindness types are given in the remaining panels. ℛ code available in Appendix 1, section A1.1.1.

Figure I.1 shows the colors used in this book; they were choosen to be discriminated by color blind people. Most figures should be understandable in black and white because the illustrative variables are also encoded by the shape of points. For figures that are in color, a link is provided that leads to a color version. Here is a brief description of these variables:

– *Leading*: used for a coding sequence, which is transcribed divergently from the origin of replication and then encoded in the leading strand for replication as illustrated in Figure I.4.

– *Lagging*: used for a coding sequence, which is transcribed convergently toward the origin of replication and then encoded in the lagging strand for replication. In bacteria, there is no documented example of a coding sequence being simultaneously leading and lagging[2].

– *Ribosomal*: used for a sequence coding of a ribosomal protein, which is used as a proxy for sequences with a high expressivity, that is with a high expression level in at least some environmental conditions.

– *Integral membrane protein*: used for a sequence coding of an integral membrane protein; its location in the hydrophobic phospholipid bilayer requires an enrichment in hydrophobic amino acids.

An elementary knowledge, or at least practice, of data dimension reduction methods such as principal component analysis (PCA) is assumed. Figure I.2 provides a basic introduction to PCA.

Let **C** be the set of the 64 possible codons:

$$\mathbf{C} = \{AAA, AAC, AAG, \ldots, TTT\} \qquad [I.1]$$

Let **A** be the set of the pseudo-amino acid Stp plus the 20 possible amino acids in proteins:

$$\mathbf{A} = \{Stp, Ala, Arg, \ldots, Val\} \qquad [I.2]$$

A genetic code is a surjective function from **C** onto **A**: every element of **C** maps to one element in **A**, and every element of **A** is mapped to by some elements of **C**, as in Figure I.3, corresponding to the so-called universal genetic code. Codons for the same amino acid are termed *synonymous*. Genetic codes in their standard presentation (as in Table I.1) are available at our ℝ's prompt with the tablecode() function.

---

2 Since the unique general rule in biology is that there are no general rules in biology, it is just a matter of time for such an example to be exhibited.

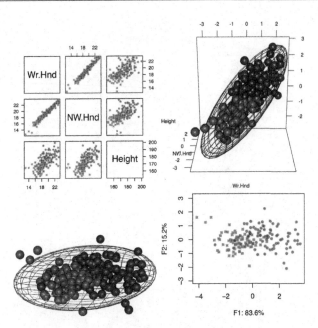

Figure I.2. *Principal component analysis in a nutshell. For a color version of this figure, see www.iste.co.uk/lobry/multivariate.zip*

NOTES ON FIGURE I.2.– This data dimension reduction method is illustrated here with a simple $\mathbb{R}^3 \rightarrow \mathbb{R}^2$ projection. The top left panel is a direct representation of the data set extracted from survey in package MASS [VEN 02]: three morphometric variables (span of writing hand, span of non-writting hand, height) for 168 students with males in blue circles and females in red squares. The top right panel is a static screenshot of the 3D representation obtained with the rgl package [ADL 14] that allows you to rotate the cloud interactively (it is worth a try!), and doing so we will probably spontaneously end with the bottom left representation in order to keep as much information as possible. The bottom right panel shows that the first factorial map obtained from PCA is the $\mathbb{R}^3 \rightarrow \mathbb{R}^2$ projection we want. ℝ code available in Appendix 1, section A1.1.2.

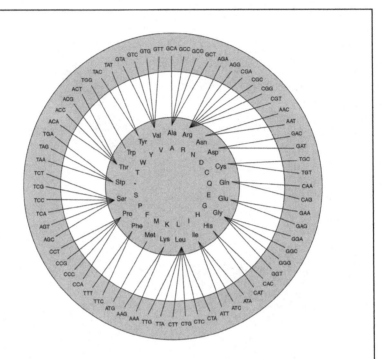

Figure I.3. *The surjective nature of genetic code. For a color version of this figure, see www.iste.co.uk/lobry/multivariate.zip*

NOTES ON FIGURE I.3.– The standard genetic code, as in Table I.1, is used here to illustrate the surjective nature of this application. The 64 codons are on the outer ring, the inner circle contains the 20 corresponding amino acids and the stop signal given with their three-letters (e.g. Gly, Ala, Stp) and one-letter (e.g. G, A, *) notations. ℝ code available in Appendix 1, section A1.1.3.

TTT Phe	TCT Ser	TAT Tyr	TGT Cys
TTC Phe	TCC Ser	TAC Tyr	TGC Cys
TTA Leu	TCA Ser	TAA Stp	TGA Stp
TTG Leu	TCG Ser	TAG Stp	TGG Trp
CTT Leu	CCT Pro	CAT His	CGT Arg
CTC Leu	CCC Pro	CAC His	CGC Arg
CTA Leu	CCA Pro	CAA Gln	CGA Arg
CTG Leu	CCG Pro	CAG Gln	CGG Arg
ATT Ile	ACT Thr	AAT Asn	AGT Ser
ATC Ile	ACC Thr	AAC Asn	AGC Ser
ATA Ile	ACA Thr	AAA Lys	AGA Arg
ATG Met	ACG Thr	AAG Lys	AGG Arg
GTT Val	GCT Ala	GAT Asp	GGT Gly
GTC Val	GCC Ala	GAC Asp	GGC Gly
GTA Val	GCA Ala	GAA Glu	GGA Gly
GTG Val	GCG Ala	GAG Glu	GGG Gly

Table I.1. *Genetic code number 1: standard*

I.2. The case under study is a genomic monster

The bacteria[3] *Borrelia burgdorferi sensu lato*, which will eventually be renamed as *Borreliella burgdorferi* [ADE 14, BAR 17], is also termed the "Lyme disease group of spirochaetes" or "Lyme disease *Borrelia*". The disease caused by this bacterium is known as borreliosis and is spread by ticks.

The complete genome of *B. burgdorferi*, sequenced in 1997 [FRA 97, CAS 00], consists of an unusual *linear* chromosome and no less than 12 linear and 9 circular plasmids containing more than 40% of the coding potential of the cell, so that it was suggested that these plasmids are in fact

---

3 In this book, the term *bacteria* is used as in HAECKEL [HAE 94], which is admittedly a little bit outdated.

minichromosomes [BAR 93]. Its chromosome is under both a strong symmetric directional mutation pressure [SUE 62, SUE 88] and a strong asymmetric directional mutation pressure [LOB 96a], but the latter is the most extreme as evidenced by many early studies on its codon and amino acid usage [MCI 98, MCL 98, PED 99, KAN 99, ROC 99b, LAF 99, MAC 99b, MAC 99c, LOB 00, TIL 00, KOW 01, ERM 01, LOP 01, KOW 02]. *B. burgdorferi* was found to have the most intense asymmetric directional mutation pressure among 43 genomes investigated [LOB 02], and in a study [JOE 18] based on 7,738 bacterial species, the *Borreliaceae* members are still the most extreme outliers. As a result of this asymmetric directional mutation pressure, the chromosome of *B. burgdorferi* has a strong chirochore structure (see Figure I.4) that was used for the first time to predict an origin of replication with a bioinformatic approach that was then experimentally confirmed [PIC 99].

The advantage of working with such a genomic monster is that many biological factors are shaping the structure of the data in an extreme way, allowing one to check the robustness of methods in a hostile environment. The disadvantage is that *B. burgdorferi* is not a typical representative of codon and amino acid usage in bacteria, at least for the relative importance of the underlying factors.

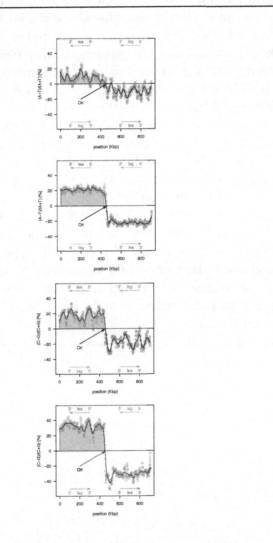

Figure I.4. *The chirochore structure of bacterial chromosomes. For a color version of this figure, see www.iste.co.uk/lobry/multivariate.zip*

NOTES ON FIGURE I.4.– Two statistics [LOB 96a] of the deviation from Parity Rule 2 (PR2) state [SUE 95] are computed: the GC-skew on the left panels and the AT-skew on the right panels using a 10-kb moving window and a 2-kb incremental step on the *B. burgdorferi* chromosome. On the top, the whole chromosome is used, while on the bottom, only third codon positions are taken into account. A chirochore, a purely descriptive notion without reference to any mechanism, is a segment homogeneous for the deviations from PR2, like the one highlighted in gray. A replichore [BLA 97] is a segment between an origin and a terminus for replication. The good thing is that their boundaries often coincide in bacteria [LOB 96a, LOB 96b, LOB 96c, FRE 98, MRÁ 98, GRI 98a, GRI 98b, MCL 98, KAR 98, MCI 98, SAL 98, CEB 98, KAR 99, LAF 99, ROC 99b, ROC 99a, LOP 99, CEB 99, MAC 99b, MAC 99a, MAC 99c, LOP 01, SER 08, XIA 12, LUO 18]. Ori points to the experimentally mapped [PIC 99] origin of replication. The arrows show the classification of coding sequences in the leading or lagging group as a function of their orientation with respect to replication. ℝ code available in Appendix 1, section A1.1.4.

Introduction to Correspondence Analysis

1.1. Chapter objectives

This multivariate data analysis technique[1] is well suited for amino acid and codon count tables. Its application, however, is not without pitfalls [PER 02] and its popularity not as high as one may expect [TEK 16]. Its primary goal is to transform a table of counts into a graphical representation, in which each gene (or protein) and each codon (or amino acid) is depicted as a point. Correspondence analysis (CA) may be defined as a special case of principal components analysis (PCA) with a different underlying metric. The purpose of this chapter is to introduce CA for someone who is already familiar with dimension reduction methods such as PCA. The underlying metric is progressively introduced and some useful properties of CA are then mentioned.

---

1 CA is also known as reciprocal averaging, method of reciprocal averages, dual scaling, canonical scoring, additive scoring, appropriate scoring, Guttman's weighting, principal component analysis of qualitative data, optimal scaling, Hayashi's theory of quantification, simultaneous linear regression and correspondence factor analysis (see [GRE 84, NIS 80] for a review). The package ade4 used here implements the duality diagram approach [DRA 07, HOL 08].

1.2. Metric choice

1.2.1. *A small data set example*

The interest of the metric in CA, that is the way we measure the distance between two individuals, is illustrated here with a very simple example, inspired by [GAU 87] and given in Figure 1.1, with only three proteins having only three amino acids, so that the consequences of the metric choice are exactly represented on a map.

```
data(toyaa)
toyaa

  Ala Val Cys
1 130  70   0
2  60  40   0
3  60  35   5
```

	$\mathbf{y}_1$	$\cdots$	$\mathbf{y}_j$	$\cdots$	$\mathbf{y}_J$	$\sum_{j=1}^{J}$
$\mathbf{x}_1$	n_{11}	$\cdots$	n_{1j}	$\cdots$	n_{1J}	$n_{1\bullet}$
$\vdots$	$\vdots$	$\ddots$	$\vdots$	$\ddots$	$\vdots$	$\vdots$
$\mathbf{x}_i$	n_{i1}	$\cdots$	n_{ij}	$\cdots$	n_{iJ}	$n_{i\bullet}$
$\vdots$	$\vdots$	$\ddots$	$\vdots$	$\ddots$	$\vdots$	$\vdots$
$\mathbf{x}_I$	n_{I1}	$\cdots$	n_{Ij}	$\cdots$	n_{IJ}	$n_{I\bullet}$
$\sum_{i=1}^{I}$	$n_{\bullet 1}$	$\cdots$	$n_{\bullet j}$	$\cdots$	$n_{\bullet J}$	$n_{\bullet\bullet}$

Table 1.1. *Contingency table*

NOTES ON TABLE 1.1.– A contingency table between two categorical variables $\mathbf{x}$ and $\mathbf{y}$. Variable $\mathbf{x}$ has I levels and $\mathbf{y}$ has J levels. The entry n_{ij} is a positive integer corresponding to the number of times level $\mathbf{x}_i$ and level $\mathbf{y}_i$ were simultaneously observed. The row marginal totals are noted $n_{i\bullet}$ and the column marginal totals $n_{\bullet j}$. The total number of observations is $n_{\bullet\bullet}$.

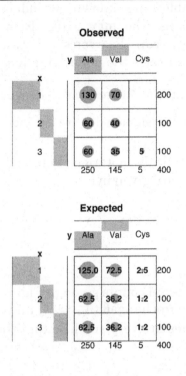

Figure 1.1. *Balloonplot. For a color version of this figure, see www.iste.co.uk/lobry/multivariate.zip*

NOTES ON FIGURE 1.1.– The balloonplot representation [WAR 16] is a graphical matrix where each cell contains a blue circle whose size reflects the relative magnitude of the corresponding component. The marginal $n_{i\bullet}$ and $n_{\bullet j}$ are graphically represented in gray behind the row or column label area. The small data set from Chapter 1 is used here. The top table contains the observed counts n_{ij}, the bottom table the expected counts n_{ij}^{t} under the hypothesis of independence given by equation [1.3]. ℝ code available in Appendix 1, section A1.2.1.

Contingency tables are known to induce a Pavolovian reflex in statisticians who may suffer from severe closure issues unless they run Pearson's χ^2 test [PEA 00] for independence. In our case, the underlying hypotheses are:

H_0:the proteins and the amino acids are independent   [1.1]

H_1:the proteins and the amino acids are not independent   [1.2]

Under H_0 and with the notations defined in Table 1.1, the expected counts n_{ij}^t are given by:

$$n_{ij}^t = n_{\bullet\bullet}P(\mathbf{x}_i \cap \mathbf{y}_j) = n_{\bullet\bullet}P(\mathbf{x}_i)P(\mathbf{y}_j) = n_{\bullet\bullet}\frac{n_{i\bullet}}{n_{\bullet\bullet}}\frac{n_{\bullet j}}{n_{\bullet\bullet}} = \frac{n_{i\bullet}n_{\bullet j}}{n_{\bullet\bullet}}$$

[1.3]

The table of expected counts given in Figure 1.1 shows that marginal totals are unchanged ($n_{i\bullet} = n_{i\bullet}^t$ and $n_{\bullet j} = n_{\bullet j}^t$). Note that some n_{ij}^t entries are very small so that the statistic, χ^2_{obs}, used in the test,

$$\chi^2_{\text{obs}} = \sum_{i=1}^{I}\sum_{j=1}^{J}\frac{\left(n_{ij} - n_{ij}^t\right)^2}{n_{ij}^t}$$

[1.4]

may poorly converge toward a χ^2 distribution. We use then a simulation approach by random sampling from the set of all contingency tables having the same marginal totals:

```
set.seed(1) # For reproducibility
chisq.test(toyaa, simulate = TRUE, B = 10^4)
```

```
        Pearson's Chi-squared test with simulated p-value (based on
            10000 replicates)
data:  toyaa
X-squared = 15.917, df = NA, p-value = 0.003
```

As illustrated by Figure 1.2, the distance defined by equation [1.4] between the table of observed counts, n_{ij}, and the table of expected counts, n^t_{ij}, is anomalously large so that we reject H_0. Figure 1.3 shows that the rejection is mainly due to the Cys amino acid, which is exclusively present in the third protein.

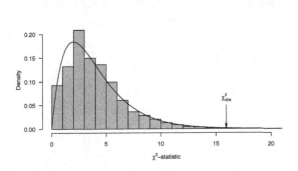

Figure 1.2. *Test for independence*

NOTES ON FIGURE 1.2.– This is an illustration of Pearson's χ^2 test [PEA 00] for independence. Under the null hypothesis H_0 given by equation [1.1], the χ^2-statistic defined by equation [1.4] is computed for 10,000 tables having the same marginal totals as in the actual data set. The histogram is then the expected distribution of this statistic under H_0. The χ^2_{obs} obtained for the actual data set using equation [1.4] is pointed out by the arrow. It is unlikely to get such a high value just by chance, so that H_0 is rejected. But note that in some rare events, representing about 0.3% of the simulated tables, a higher value is obtained. This proportion of 0.3% is called the *P*-value and is the risk associated with the decision of rejecting H_0. The continuous line is the density function for the χ^2 distribution with $(3 - 1) \times (3 - 1) = 4$ degrees of freedom. For data sets that are not too pathological, the distribution will converge toward this probability density function, allowing the simulation approach to be avoided. ⍴ code available in Appendix 1, section A1.2.2.

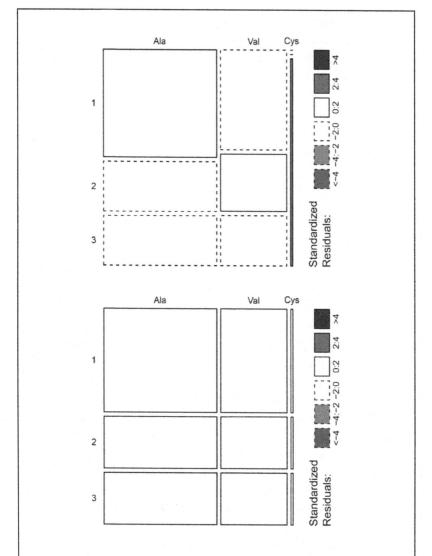

Figure 1.3. *Mosaicplot. For a color version of this figure, see www.iste.co.uk/lobry/multivariate.zip*

NOTES ON FIGURE 1.3.– A mosaicplot [HAR 84, EME 98, FRI 94] of the small data set. The area of the rectangles is proportional to the observed counts n_{ij} in the top panel and to the expected counts n_{ij}^t in the bottom panel, as shown in Figure 1.1. The color code helps to understand how the actual data differs from what is expected under H_0, here in blue the excess of the amino acid Cys in the third protein. This representation works well only with variables with a limited number of levels. ℜ code available in Appendix 1, section A1.2.3.

1.2.2. Euclidean distance

The Euclidean distance between two proteins $\mathbf{x}_i$ and $\mathbf{x}_{i'}$ is defined, with the notations used in Table 1.1, by:

$$d^2(\mathbf{x}_i, \mathbf{x}_{i'}) = \sum_{j=1}^{J} \left(n_{ij} - n_{i'j} \right)^2 \qquad [1.5]$$

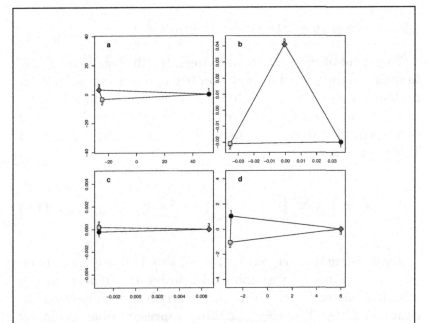

Figure 1.4. *Take home message. For a color version of this figure, see www.iste.co.uk/lobry/multivariate.zip*

NOTES ON FIGURE 1.4.– Do not mess with the metric choice: the distance between the three proteins in the small data set is represented using the Euclidean distance 1.5 in panel **a**, the Euclidean distance on protein profiles 1.6 in panel **b**, the Euclidean distance on double profiles 1.7 in panel **c** and the χ^2 distance 1.8 in panel **d**. ® code available in Appendix 1, section A1.2.4.

The corresponding graphical representation is given by panel **a** in Figure 1.4. From this point of view, the first protein is far away from the two others. But thinking about it, this is a rather trivial effect of protein size if we consider the raw marginal totals shown in Figure 1.1. With 200 amino acids, the first protein is two times bigger than the others so that when computing the Euclidean distance [1.5] its n_{ij} entries are on average bigger, sending it far away.

1.2.3. *Euclidean distance on row profiles*

To get rid of the protein size effect, the first obvious idea is to divide counts by the total number of amino acids in each protein so as to work with vectors of amino acid relative frequencies, which are called *protein profiles*. The corresponding distance, using the notations given in Table 1.2, is given by:

$$d^2(\mathbf{x}_i, \mathbf{x}_{i'}) = \sum_{j=1}^{J} \left(\frac{n_{ij}}{n_{i\bullet}} - \frac{n_{i'j}}{n_{i'\bullet}} \right)^2 = \sum_{j=1}^{J} \left(r_{ij} - r_{i'j} \right)^2 \qquad [1.6]$$

By construction we have $r_{i\bullet} = 1$, this is a *protein profile*: we are working with amino acid relative frequencies in the proteins instead of amino acid absolute frequencies in equation [1.5]. The corresponding representation given by panel **b** in Figure 1.4 is now completely different with the three proteins equally spaced. This is because, in terms of relative amino acid composition, they all differ two-by-two by 5% at the level of two amino acids only. We have removed the trivial protein size effect, but this is still not completely satisfactory. Consider the total number of amino acids in Figure 1.1, the proteins differ by 5% for all amino acids but the situation is somewhat different for Cys because this amino acid is very rare. A difference of 5% for a rare amino acid does not have the same significance than a difference of 5% for a common amino acid such as Ala in our example.

	$\mathbf{y}_1$	$\cdots$	$\mathbf{y}_j$	$\cdots$	$\mathbf{y}_J$	$\sum_{j=1}^{J}$
$\mathbf{x}_1$	$r_{11} = \frac{n_{11}}{n_{1\bullet}}$	$\cdots$	$r_{1j} = \frac{n_{1j}}{n_{1\bullet}}$	$\cdots$	$r_{1J} = \frac{n_{1J}}{n_{1\bullet}}$	$r_{1\bullet} = \frac{n_{1\bullet}}{n_{1\bullet}} = 1$
$\vdots$	$\vdots$	$\ddots$	$\vdots$	$\ddots$	$\vdots$	$\vdots$
$\mathbf{x}_i$	$r_{i1} = \frac{n_{i1}}{n_{i\bullet}}$	$\cdots$	$r_{ij} = \frac{n_{ij}}{n_{i\bullet}}$	$\cdots$	$r_{iJ} = \frac{n_{iJ}}{n_{i\bullet}}$	$r_{i\bullet} = \frac{n_{i\bullet}}{n_{i\bullet}} = 1$
$\vdots$	$\vdots$	$\ddots$	$\vdots$	$\ddots$	$\vdots$	$\vdots$
$\mathbf{x}_I$	$r_{I1} = \frac{n_{I1}}{n_{I\bullet}}$	$\cdots$	$r_{Ij} = \frac{n_{Ij}}{n_{I\bullet}}$	$\cdots$	$r_{IJ} = \frac{n_{IJ}}{n_{I\bullet}}$	$r_{I\bullet} = \frac{n_{I\bullet}}{n_{I\bullet}} = 1$
$\sum_{i=1}^{I}$	$r_{\bullet 1} = \frac{n_{\bullet 1}}{n_{\bullet\bullet}}$	$\cdots$	$r_{\bullet j} = \frac{n_{\bullet j}}{n_{\bullet\bullet}}$	$\cdots$	$r_{\bullet J} = \frac{n_{\bullet J}}{n_{\bullet\bullet}}$	$r_{\bullet\bullet} = \frac{n_{\bullet\bullet}}{n_{\bullet\bullet}} = 1$ I

Table 1.2. *The table of row profiles r_{ij}.*

1.2.4. *Euclidean distance on double profiles*

Let us use a similar but transposed strategy using a *double profile* by dividing simple profile values by column marginal totals:

$$d^2(\mathbf{x}_i, \mathbf{x}_{i'}) = \sum_{j=1}^{J} \left(\frac{r_{ij}}{n_{\bullet j}} - \frac{r_{i'j}}{n_{\bullet j}} \right)^2 = \sum_{j=1}^{J} \frac{1}{n_{\bullet j}^2} \left(r_{ij} - r_{i'j} \right)^2 \quad [1.7]$$

It is interesting to note that this distance on double profile is readable as a *weighted* Euclidean distance on protein profiles: the terms in equation [1.6] are weighted by $1/n_{\bullet j}^2$. Panel **c** in Figure 1.4 shows that now protein 3 is sent far away because it is very rich in the very rare amino acid Cys. Hence, with this double profile we have removed, first, the protein size effect by dividing by marginal row counts and, second, the global amino acid frequency effect by dividing by marginal column profile.

1.2.5. The χ^2 distance

The distance used in CA is similar to double profile distance 1.7 but with a different weighting scheme:

$$d^2(\mathbf{x}_i, \mathbf{x}_{i'}) = n_{\bullet\bullet} \sum_{j=1}^{J} \left(\frac{r_{ij}}{\sqrt{n_{\bullet j}}} - \frac{r_{i'j}}{\sqrt{n_{\bullet j}}} \right)^2 = n_{\bullet\bullet} \sum_{j=1}^{J} \frac{1}{n_{\bullet j}} (r_{ij} - r_{i'j})^2$$

[1.8]

This distance is understandable as a *weighted* Euclidean distance on protein profiles: the terms in equation [1.6] are weighted by $1/n_{\bullet j}$. Panel **d** in Figure 1.4 shows that protein 3 is sent away but less than in the double profile approach: the coefficient $1/n_{\bullet j}$ in equation [1.8] gives less weight than the coefficient $1/n_{\bullet j}^2$ in equation [1.7] to the rare Cys amino acid. The distance defined by equation [1.8] is also known as the χ^2 metric. This is related to the independence test based on contingency tables. Under H_0 all proteins share a common profile given by the column marginal totals $r_{\bullet j}$, we can then compute the distance 1.8 between any protein $\mathbf{x}_i$ and this profile:

$$d^2(\mathbf{x}_i, r_{\bullet j}) = n_{\bullet\bullet} \sum_{j=1}^{J} \frac{1}{n_{\bullet j}} (r_{ij} - r_{\bullet j})^2$$

[1.9]

To get an overall view of the departure from H_0, we can compute $\bar{g}$, the average distance among all proteins, weighting their contribution as $n_{i\bullet}/n_{\bullet\bullet}$, the fraction of the total number of amino acids they account for:

$$\bar{g} = \sum_{i=1}^{I} \frac{n_{i\bullet}}{n_{\bullet\bullet}} d^2 \left(\mathbf{x}_i, r_{\bullet j} \right) \tag{1.10}$$

$$= \sum_{i=1}^{I} \frac{n_{i\bullet}}{n_{\bullet\bullet}} n_{\bullet\bullet} \sum_{j=1}^{J} \frac{1}{n_{\bullet j}} \left(r_{ij} - r_{\bullet j} \right)^2 \tag{1.11}$$

$$= \sum_{i=1}^{I} \sum_{j=1}^{J} \frac{n_{i\bullet}}{n_{\bullet j}} \left(\frac{n_{ij}}{n_{i\bullet}} - \frac{n_{\bullet j}}{n_{\bullet\bullet}} \right)^2 \tag{1.12}$$

$$= \sum_{i=1}^{I} \sum_{j=1}^{J} \frac{n_{i\bullet}}{n_{\bullet j} n_{i\bullet}^2} \left(n_{ij} - \frac{n_{i\bullet} n_{\bullet j}}{n_{\bullet\bullet}} \right)^2 \tag{1.13}$$

$$= \sum_{i=1}^{I} \sum_{j=1}^{J} \frac{1}{n_{i\bullet} n_{\bullet j}} \left(n_{ij} - \frac{n_{i\bullet} n_{\bullet j}}{n_{\bullet\bullet}} \right)^2 \tag{1.14}$$

$$= \sum_{i=1}^{I} \sum_{j=1}^{J} \frac{1}{n_{i\bullet} n_{\bullet j}} \left(n_{ij} - n_{ij}^t \right)^2 \tag{1.15}$$

$$= \sum_{i=1}^{I} \sum_{j=1}^{J} \frac{1}{n_{\bullet\bullet}} \frac{n_{\bullet\bullet}}{n_{i\bullet} n_{\bullet j}} \left(n_{ij} - n_{ij}^t \right)^2 \tag{1.16}$$

$$= \sum_{i=1}^{I} \sum_{j=1}^{J} \frac{1}{n_{\bullet\bullet}} \frac{\left(n_{ij} - n_{ij}^t \right)^2}{n_{ij}^t} \tag{1.17}$$

$$= \frac{\chi_{\text{obs}}^2}{n_{\bullet\bullet}} \tag{1.18}$$

When using distance 1.8 to compute the average distance $\bar{g}$ between the proteins and the expected protein under H_0, the result is proportional to the statistic 1.4 used in Pearson's χ^2 test for independence. We can check this numerically in the following way:

```
all.equal(as.numeric(chisq.test(toyaa, simulate = TRUE) $stat),
          sum(toyaa)*sum(dudi.coa(toyaa, scan=F) $eig))
```

[1] TRUE

Because the expected protein under H_0 is also at the gravity center of proteins, $\bar{g}$ is also referred to as the *total inertia* or the *total variability*.

1.3. Properties

1.3.1. *Scree plot of eigenvalues*

When the null hypothesis 1.1 for Pearson's χ^2 test [PEA 00] for independence is rejected, we know that the two variables are not independent, but we do not know why. The aim of CA, similar to the mosaicplot [HAR 84, EME 98, FRI 94] in Figure 1.3, is to help understand why. CA yields an orthogonal system of axes usually called factors and denoted as F_1, F_2, F_3, ... The first factor corresponds to the largest variability, the second to the second largest variability and which is orthogonal to the first factor and so on. The variability accounted by factor F_i is given by its associated eigenvalue λ_i. The scree plot [CAT 66] graphs the eigenvalues against the factor numbers. Inspection of the scree plot can be very important for the choice of the number of factors to be retained. For instance, if the second and third were very close and the user chose to take two axes, this could result in wrong results [HOL 85]. Figure 1.5 illustrates what happens when two eigenvalues are becoming too close: the plane is stable but not each individual factor.

Save for the general advice to avoid cutting between two eigenvalues that are too close [HOL 85], there is no magic algorithm to decide for you the number of factors to keep, only numerous guidelines [PER 05]. Note that ® provides the r2dtable() function that implements PATEFIELD's algorithm [PAT 81] to generate random contingency tables with given marginal totals. It is therefore straightforward by simulations to have an idea of the distribution of eigenvalues in the scree plot under the null hypothesis 1.1, as shown in Figure 1.6. This approach is generically termed the *parallel analysis* in [HOW 16].

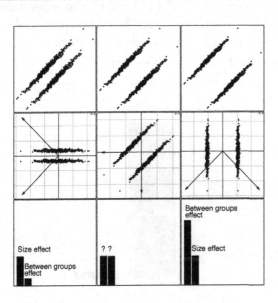

Figure 1.5. *Degeneracy*

NOTES ON FIGURE 1.5.– From top to bottom, we have the original data set, the first factorial map and the scree plot for eigenvalues. There are two main sources of variability: a strong correlation between the two variables, often referred to as a size effect in multivariate analysis when there are many positively correlated variables, and a shift between two groups. From left to right, the between-group variability is increased. In the left column, the between-groups effect is small as compared to the size effet as illustrated by the eigenvalue graph. The two groups are separated by the second factor in the factorial map. In the right column, we have the opposite situation, the between-groups effect is big as compared to the size effect as illustrated again by the eigenvalue graph. The two groups are separated by the first factor in the factorial map. The middle column is a degenerate case in which the between-groups effect and the size effet are very close, there is no way to order these effects so that the eigenvalues are the same. Any rotation of the factorial map with respect to the origin would summarize the data as well. ℝ code available in Appendix 1, section A1.2.5.

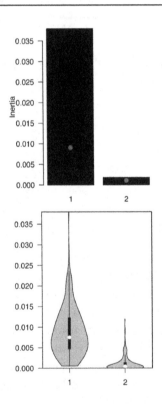

Figure 1.6. *Scree plot for the small data set. For a color version of this figure, see www.iste.co.uk/lobry/multivariate.zip*

NOTES ON FIGURE 1.6.– Scree plot of the eigenvalues for correspondence analysis of the small protein data set with, at the top, the actual data, and, underneath, 250 simulations under the null hypothesis 1.1. The red points are the mean values under the independency hypothesis, and the white points are the median values. ℝ code available in Appendix 1, section A1.2.6.

```
set.seed(123)
df <- data.frame(1:5, rnorm(5))
prcomp(df)$x


           PC1         PC2
[1,] -2.1144924 -0.3122616
[2,] -1.0671658 -0.2017905
[3,]  0.2898160  1.3340197
[4,]  0.9510792 -0.3325546
[5,]  1.9407629 -0.4874130


princomp(df)$scores


         Comp.1      Comp.2
[1,]  2.1144924  0.3122616
[2,]  1.0671658  0.2017905
[3,] -0.2898160 -1.3340197
[4,] -0.9510792  0.3325546
[5,] -1.9407629  0.4874130
```

1.3.2. *Factor orientations are meaningless*

From the distance between the three proteins, there are countless ways of plotting them on a map but fundamentally four symmetrical solutions that are not equivalent by rotation (see Figure 1.7). From an algebraic point of view, if $\mathbf{v}$ is an eigenvector of matrix $\mathbf{M}$ with eigenvalue λ, then its opposite, $-\mathbf{v}$, is an eigenvector too:

$$\mathbf{Mv} = \lambda \mathbf{v} \iff -\mathbf{Mv} = -\lambda \mathbf{v} \iff \mathbf{M}(-\mathbf{v}) = \lambda(-\mathbf{v}) \quad [1.19]$$

The sign of the eigenvector returned by algorithms to compute them is arbitrary. For instance, the function prcomp() and princomp() are based on two different algorithms and could differ in the sign of the results:

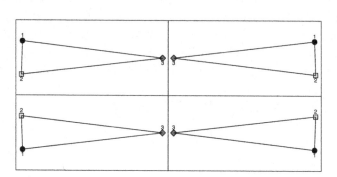

Figure 1.7. *Symetries. For a color version of this figure, see www.iste.co.uk/lobry/multivariate.zip*

NOTES ON FIGURE 1.7.– For a given distance definition, here CA, there are four different ways to locate the three proteins on a map. The figures on the left are symmetrical with respect to a vertical line to those on the right. The figures on the top are symmetrical with respect to a horizontal line to those on the bottom. ℝ code available in Appendix 1, section A1.2.7.

Because of the deterministic behavior of computers, a given algorithm on a given platform on a given data set will of course always return the same result, but just changing the platform may flip some factors[2]. The factor *orientations* are meaningless, no interpretation should be deduced from them.

---

2 I have noted with great interest from ℝ-devel NEWS when version 3.4.4 was already released (2018-03-15) that "[b]y default the (arbitrary) signs of the loadings from princomp() are chosen so the first element is non-negative". Since ℝ is the *lingua franca* for statistics, we may have an emerging standard here that will save a lot of time for professors with students who feel reluctant toward the understanding of elementary symmetry invariants.

1.3.3. *CA is symmetric*

An important property of CA is its perfect symmetry with respect to the rows and columns: exactly the same results are obtained when running the analysis on the original table or its transposed version as illustrated by Figure 1.8. This allows for the *simultaneous* representation of the levels of the variable in row and the levels of the variable in column. As a result, the interpretation of factorial maps in CA is very intuitive. In our example, the main factor of variability is the opposition between protein 3 and proteins 1 and 2, with protein 3 enriched in the amino acid Cys; the residual variability is the opposition between protein 1 and protein 2, with protein 1 enriched in the amino acid Ala and protein 2 enriched in the amino acid Val. There is a *correspondence* between protein 3 and the amino acid Cys, which in our case is extreme because Cys is exclusively found in protein 3.

1.3.4. *Nonlinearity in CA is not a bug but a feature*

Nonlinearity in CA factorial maps, more exactly a parabolic-shaped cloud of points, was called *effet Guttman* by BENZÉCRI [BEN 73] but is also known as the circumplex in psychology, the horseshoe in archaeology and the arch in ecology [WAR 87]. Since there are no nicknames in codon usage studies, and as a mnemonic tip, I will call this the rainbow. Suppose that there is a hypothetical gradient of amino acid concentration in which Ala is progressively replaced by Val and then by Cys as illustrated by panel **a** in Figure 1.9. The proteins are regularly placed along this gradient as illustrated by the red ticks on the x-axis, and their amino acid content directly proportional to the amino acid concentration at their location. We have then *lowphiles* proteins enriched in Ala, *highphiles* proteins enriched in Cys and the *middlephiles* proteins enriched in Val but not completely depleted from Ala and Cys. The direct representation of proteins counts in panel **b** shows that proteins are on a loop starting from *lowphiles* close to the origin up to the *highphiles* close again to the origin through the *middlephiles* away from the origin. A representation of the protein profiles is given by the triangular plot in panel **c**, the proteins are along an arch. The shape of points is intrinsically curved: this is a direct representation of protein profiles, without projection or distortion. This curvature is preserved in the first factorial map of CA given in panel **d**, the first factor of variability, F_1, is the opposition between the *lowphiles* and the *highphiles*, the second factor of variability, F_2, is the opposition between the *lowphiles* and all the others proteins. Nonlinearity in CA is not a bug but a feature [WAR 87]. When the major factor of variability in your data is an underlying gradient, as the colors in a rainbow, our first factorial map will look like a rainbow.

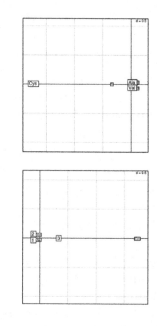

Figure 1.8. *Transposition invariance. For a color version of this figure, see www.iste.co.uk/lobry/multivariate.zip*

NOTES ON FIGURE 1.8.– The first factorial map of CA for the original data, at the top, is exactly the same (neglecting symmetries as depicted in Figure 1.7) as the one obtained, underneath, with the transposed data set. In these representations obtained with the function scatter() from package ade4, the size of character is bigger for the levels of the variable in row. ℝ code available in Appendix 1, section A1.2.8.

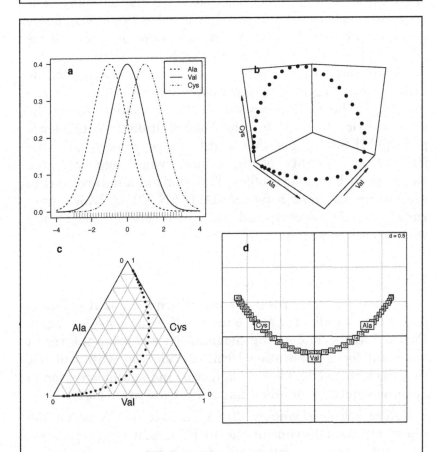

Figure 1.9. *Rainbow effect*

NOTES ON FIGURE 1.9.– See text in section 1.3.4 for explanations. Panel **a** shows the gradient simulation,

> **b** the 3D representation, **c** the triangular plot and **d** the first factorial map. ℝ code available in Appendix 1, section A1.2.9.

1.3.5. *Distributional equivalence principle*

Suppose that two proteins x_i and $x_{i'}$ share the same profile, then the χ^2 distance 1.8 between them is zero: they are located at the same position on factorial maps. The distributional equivalence principle ensures that if you merge the two proteins x_i and $x_{i'}$ into a new one, x_{i*} where $n_{i*j} = n_{ij} + n_{i'j}$, in the contingency table, and delete originals, factorials maps are unaltered. Because CA is symmetric, the distributional equivalence principle is also true for the columns of the contingency table. The χ^2 distance is not the sole metric to satisfy the distributional equivalence principle [ESC 78], but this is still a nice property allowing for a great stability of the results when agglomerating elements with similar profiles. This leads in a natural way to the within CA and between CA [BEN 83] in which total variability is decomposed in its between-block and within-block components.

1.3.6. *So, why the χ^2 metric?*

To quote MARX[3], "Those are my principles, and if you don't like them... well, I have others". If you consider again Figure 1.4, the underlying meaning is: whatever the result you want for a publication I can provide you with a metric that fits your needs. There is in fact no indisputable reason to favor one metric over other one, so here are my € 0.02.

– *Rare allele advantage*: the χ^2 metric in CA is not the commonly used Euclidean one in PCA, and since scientists are easily bored we will benefit from a frequency-dependent selective process.

------

3 Groucho, not Karl.

– *Phylogenetic inertia*: according to KUHN [KUH 62], a paradigm shift lasts about 40 years (two scientific generations) but the χ^2 metric is still in use 120 years after its introduction [PEA 00]. It would be foolish to fight against phylogenetic inertia, just think about the very under-optimal location of keys in QWERTY keyboards. We may also think of this as a backward compatibility argument.

– *Convergent evolution*: CA has been re-invented many times in very different cultural environments (see footnote 1), so that it is hard to believe that these bright scientists were all wrong.

– *Founder effect*: the first codon usage studies [GRA 80b, GRA 80a] were based on the χ^2 metric, so working with it make us feel that we are sitting on the shoulders of giants.

– *Fitness landscape stability*: in the context of an exponentially growing database, I have always been fascinated by the stability of results over years as illustrated by Figure 1.10. For bacterial complete genomes, there is an extreme taxonomic sampling bias in favor of species related to human health (e.g. the 3,239 strains from *Staphylococcus aureus* represent 25% of available data). Factors in CA are insensitive to these taxonomic sampling biases: this is partly due to the low within-species variability for codon usage but also partly to the distributional equivalence principle. As compared to [LOB 03], the size of the data set is increased by two orders of magnitude, as shown in Figure 1.10, and the structure is unaltered.

– *Dyslexia friendly*: even if we remember properly what i and j stands for in n_{ij}, matrices are filled by rows in the mathematical world and by columns in the statistical world, which is an endless source of headaches. Since CA is symmetric we will save a lot of acetyl-salicylic acid.

– *Tribute to neutral evolution*: as already pointed out at the turn of the millennium [GAU 00], SUEOKA was the first to state in 1962, before the emergence of neutralism, that some patterns at the genomic level could appear without natural selection [SUE 62]. Figure 1.11 shows that CA captures spontaneously the spirit of SUEOKA's neutrality plot [SUE 88].

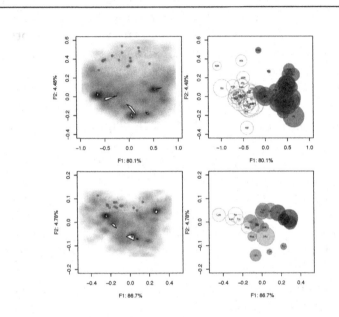

Figure 1.10. *Stability of results over time. For a color version of this figure, see www.iste.co.uk/lobry/multivariate.zip*

NOTES ON FIGURE 1.10.– The first factorial map for codon usage in bacteria with synonymous usage in top panels and non-synonymous usage in bottom panels. The left panels are in row space (coding sequences or proteins), and the right panels are in column space (codons or amino acids). As compared to [LOB 06], the present data set contains a total of 13,165,776,353 codons ($\times$24) from 12,317 bacterial strains ($\times$17) from 2,301 species ($\times$5) from 980 genera ($\times$4.5) but the structure is still the same. Factor F_1 is a GC-content gradient as seen in codon space with GC-ending codons in gray and in amino acid space with GC favored amino acids as defined in section 4.3.2.1 in gray. Factor F_2 is a thermophilic-related factor as seen in row space with hyperthermophilic species from [LOB 06] given by the red circles. The genome hypothesis [GRA 80b, GRA 80a] is illustrated here with the white polygons, which are convex hulls for the 3,239 strains from *Staphylococcus aureus*, the 315 strains from *Enterococcus faecalis*, the 883 strains from *Escherichia coli*, the 272 strains from *Salmonella enterica*, the 225 strains from *Klebsiella pneumoniae* and the 1,537 strains from *Mycobacterium tuberculosis*. Data were retrieved from release 7 of hogenom [PEN 09]. ℜ code available in Appendix 1, section A1.2.10.

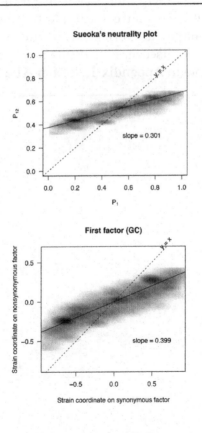

Figure 1.11. *Neutrality plot*

NOTES ON FIGURE 1.11.– SUEOKA's neutrality plot [SUE 88] is the GC content in first and second positions, P_{12}, versus the GC content in third codon positions, P_3, with a data censorship of eight codons, computed here from the complete genome of 12,317 bacterial strains. If there was no selection on the average amino acid content of proteins, points should lie on the $y = x$ line; if there was an absolute selection, points should lie on a horizontal line; while observed slope value (from orthogonal regression) is intermediate between these two extreme theoretical situations. Because the first factor for codon usage in bacteria is the GC content (see Figure 1.10), plotting in the bottom panel non-synonymous versus synonymous coordinates spontaneously rediscovers SUEOKA's neutrality plot. ℝ code available in Appendix 1, section A1.2.11.

Global Correspondence Analysis

2.1. Data set

2.1.1. *Queries in database*

We now open the database `emglib` [PER 00] that contains the extra information we are interested in (see Figure 2.1). We query[1] for all complete (no k=partial) coding sequences (t=cds) from the *Borrelia burgdorferi* complete genome (n=BORBUCG) and store the result in the object borre.

```
if(final.query1){
  choosebank("emglib")
  borre <- query("borre","n=BORBUCG and t=cds and no k=partial")
}
```

We loop over the sequences to compute, with the utility function uco(), the table tuco of codon usage, with the nseq coding sequences in rows and the 64 codons in columns. The entry tuco[i,j] in this table is a positive integer, giving for the sequence i the total number of codons of type j. We save

---

1 The ACNUC query language [GAU 82a, GAU 82b, GOU 84, GOU 85a, GOU 85b, GOU 08] is described online at http://doua.prabi.fr/databases/acnuc_data/cfonctions#QUERYLANGUAGE. Examples under ℝ are given in chapter 4 of the seqinr manual at http://seqinr.r-forge.r-project.org/.

the object `tuco` in a file using the standard `save()` function. It is worth noting that ℜ uses the XDR [SUN 87] representation of objects in binary saved files, and these are portable across all ℜ platforms. The `save()` and `load()` functions are very efficient (because of their binary nature) for saving and restoring any kind of objects, and are additionally platform-independent.

Figure 2.1. *EMGLib*

NOTES ON FIGURE 2.1.– Screenshot of the entry for the complete genome of *Borrelia burgdorferi* in EMGLib [PER 00]. As compared with general databases, we have here extra annotations. The orientation of genes with respect to replication (leading or lagging) is given under the `/strand=` qualifier, and the value of the codon adaptation index [SHA 87] is given under the `/CAI=` qualifier.

```
if(final.query1){
  nseq <- borre$nelem
  tuco <- matrix(NA, nrow = nseq, ncol = 64)
  rownames(tuco) <- getName(borre)
  colnames(tuco) <- words()
  for(i in 1:nseq) tuco[i, ] <- uco(getSequence(borre$req[[i]]))
  save(tuco, file = "local/tuco.rda")
} else {
  load("local/tuco.rda")
}


tuco[1:5, 1:15]
```

	aaa	aac	aag	aat	aca	acc	acg	act	aga	agc	agg	agt	ata	atc	atg
BORBUCG.BB0001	13	1	2	12	5	3	0	11	1	2	0	2	15	5	2
BORBUCG.BB0002	20	1	6	27	8	1	0	5	8	3	2	8	11	2	12
BORBUCG.BB0003	34	3	9	32	4	0	3	6	7	3	3	6	12	3	9
BORBUCG.BB0004	60	19	14	44	16	2	0	7	13	8	2	4	39	9	7
BORBUCG.BB0005	25	2	11	20	7	0	1	5	5	6	7	8	11	1	9

```
dim(tuco)
```

```
[1] 850  64
```

```
sum(tuco)
```

```
[1] 284122
```

Our data set is therefore a contingency table in which 284,122 codons are distributed among the crossed levels between the 64 codons and the 850 coding sequences. A global overview of the data set is shown in Figure 2.3. *B. burgdorferi* is an AT-rich bacteria (see Figure 2.2), so that without exception, AT-ending codons are always more frequent among synonymous codons.

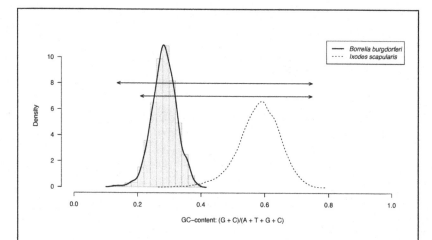

Figure 2.2. *GC content in CDS*

NOTES ON FIGURE 2.2.– The distribution on the left is the GC content in 850 coding sequences from *B. burgdorferi*. The top arrow is the observed range of genomic GC content in bacteria, ranging from 13.5% in *Candidatus Zinderia insecticola* to 74.9% in *Anaeromyxobacter dehalogenans*, out of a total of 2,670 genomes [ZHO 14]. The bottom arrow is the observed range for the average GC content in coding sequences, ranging from 20.8% in *Candidatus Sulcia muelleri* to 74.8% in *Anaeromyxobacter dehalogenans*, from the data in Figure 1.10. The distribution on the right is the GC content in 14,723 coding sequences from *Ixodes scapularis*, the best documented host tick for *B. burgdorferi*. The ℝ code for this figure is available in Appendix 1, section A1.3.1.

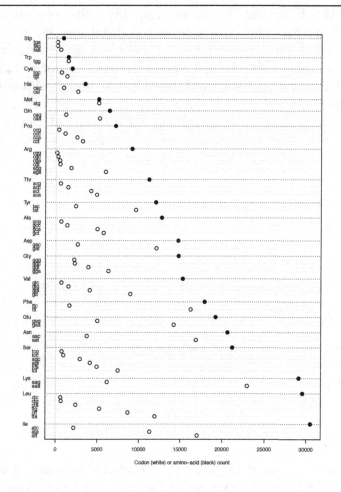

Figure 2.3. *Overview of global codon usage*

NOTES ON FIGURE 2.3.– CLEVELAND's dot plot [CLE 85] of codon (white circles) and amino acid (black circles) usage in the coding sequences from *B. burgdorferi* is shown. The ⓡ code for this figure is available in Appendix 1, section A1.3.2.

In the following, we again loop over the sequences to extract useful information from the database. We will use them as illustrative variables to help the interpretation of factorial maps. The logical vector leading is such that its element leading[i] is true if the coding sequence number i is on the leading strand with respect to replication. The numerical vector gc contains the GC-content of the coding sequences. The numerical vector cai contains the codon adaptation index (CAI) [SHA 87] value.

```
if(final.query1){
  n <- borre$nelem # total number of sequences
  leading <- logical(n)
  gc <- numeric(n)
  verbose <- TRUE
  for(i in 1:n){
    if(verbose) print(paste("Dealing with sequence number:", i))
    annot <- getAnnot(borre$req[[i]], nbl = 8)
    if(length(grep("leading", annot)) == 1) leading[i] <- TRUE
    myseq <- getSequence(borre$req[[i]])
    gc[i] <- GC(myseq)
    cai[i] <- as.numeric(unlist(strsplit(annot[3], split = '"'))[2])
  }
  save(leading, file = "local/leading.rda")
  save(gc, file = "local/gc.rda")
  save(cai, file = "local/cai.rda")
} else {
  load("local/leading.rda")
  load("local/gc.rda")
  load("local/cai.rda")
}
```

Ribosomal proteins account for about 50% of bacterial biomass during the exponential phase of growth. They are thus commonly used in codon usage studies as a proxy for sequences with a high expressivity, that is, sequences with a high expression level in at least some environmental conditions. The criterion k=@ribosomal@protein@ enables the selection in the list of only the complete coding sequences that are associated with the keyword @ribosomal@protein@ where the character @ means any string of characters.

```
if(final.query1){
  rib <- query("rib", "borre and k=@ribosomal@protein@")
  ribnames <- getName(rib)
  writeLines(ribnames, con = "local/ribnames.txt")
  closebank()
} else {
  ribnames <- readLines("local/ribnames.txt")
}
```

2.1.2. *Data polishing*

The subset of ribosomal proteins is based on a query by keyword matching, so that a false positive such as ribosomal protein L11 methyltransferase is always possible. We want to check that there are no intruders by checking the annotations of the sequences. Unfortunately, we closed the connection to the database at the end of section 2.1.1. The good thing is that the function crelistfromclientdata(), or its short form, clfcd(), allows us to make a list on the server from previously saved sequences names. We do not have to then repeat the whole query process.

```
if(final.query2){
  choosebank("emglib")
  rib <- clfcd("rib", "local/ribnames.txt", type = "SQ")
  n <- rib$nelem
  riba <- character(n)
  for(i in 1:n){
    annot <- getAnnot(rib$req[[i]], nbl = 10)
    riba[i] <- annot[grep("/product=", annot)]
    riba[i] <- unlist(strsplit(riba[i], split = '"'))[2]
  }
  closebank()
  writeLines(riba, con = "local/riba.txt")
} else {
  riba <- readLines("local/riba.txt")
}
```

Now using the package xtable, it is easy to generate the LaTeX Table 2.1 showing that there are no unwanted sequences in our ribosomal subset. The code that follows the table is used to generate it.

	Sequence name	Product
1	BB0112	Ribosomal protein L9 (rplI)
2	BB0113	Ribosomal protein S18 (rpsR)
3	BB0115	Ribosomal protein S6 (rpsF)
4	BB0123	Ribosomal protein S2 (rpsB)
5	BB0127	Ribosomal protein S1 (rpsA)
6	BB0188	Ribosomal protein L20 (rplT)
7	BB0189	Ribosomal protein L35 (rpmI)
8	BB0229	Ribosomal protein L31 (rpmE)
9	BB0233	Ribosomal protein S20 (rpsT)
10	BB0256	Ribosomal protein S21 (rpsU)
11	BB0338	Ribosomal protein S9 (rpsI)
12	BB0339	Ribosomal protein L13 (rplM)
13	BB0350	Ribosomal protein L28 (rpmB)
14	BB0386	Ribosomal protein S7 (rpsG)
15	BB0387	Ribosomal protein S12 (rpsL)
16	BB0390	Ribosomal protein L7/L12 (rplL)
17	BB0391	Ribosomal protein L10 (rplJ)
18	BB0392	Ribosomal protein L1 (rplA)
19	BB0393	Ribosomal protein L11 (rplK)
20	BB0396	Ribosomal protein L33 (rpmG)
21	BB0440	Ribosomal protein L34 (rpmH)
22	BB0477	Ribosomal protein S10 (rpsJ)
23	BB0478	Ribosomal protein L3 (rplC)
24	BB0479	Ribosomal protein L4 (rplD)
25	BB0480	Ribosomal protein L23 (rplW)
26	BB0481	Ribosomal protein L2 (rplB)
27	BB0482	Ribosomal protein S19 (rpsS)
28	BB0483	Ribosomal protein L22 (rplV)
29	BB0484	Ribosomal protein S3 (rpsC)
30	BB0485	Ribosomal protein L16 (rplP)
31	BB0486	Ribosomal protein L29 (rpmC)
32	BB0487	Ribosomal protein S17 (rpsQ)
33	BB0488	Ribosomal protein L14 (rplN)
34	BB0489	Ribosomal protein L24 (rplX)
35	BB0490	Ribosomal protein L5 (rplE)
36	BB0491	Ribosomal protein S14 (rpsN)
37	BB0492	Ribosomal protein S8 (rpsH)
38	BB0493	Ribosomal protein L6 (rplF)
39	BB0494	Ribosomal protein L18 (rplR)
40	BB0495	Ribosomal protein S5 (rpsE)
41	BB0496	Ribosomal protein L30 (rpmD)
42	BB0497	Ribosomal protein L15 (rplO)
43	BB0499	Ribosomal protein L36 (rpmJ)
44	BB0500	Ribosomal protein S13 (rpsM)
45	BB0501	Ribosomal protein S11 (rpsK)
46	BB0503	Ribosomal protein L17 (rplQ)
47	BB0615	Ribosomal protein S4 (rpsD)
48	BB0695	Ribosomal protein S16 (rpsP)
49	BB0699	Ribosomal protein L19 (rplS)
50	BB0703	Ribosomal protein L32 (rpmF)
51	BB0778	Ribosomal protein L21 (rplU)
52	BB0780	Ribosomal protein L27 (rpmA)
53	BB0804	Ribosomal protein S15 (rpsO)

Table 2.1. *List of sequences in the ribosomal subset*

```
library(xtable)
ribnames <- substring(readLines("local/ribnames.txt"), 9, 14)
rbat <- cbind(ribnames, riba)
colnames(rbat) <- c("Sequence name", "Product")
print(xtable(rbat, caption = "List of sequences in the ribosomal subset.",
    label = "riba"), file = "tables/riba.tex", size = "scriptsize")
```

A minimum threshold for the coding sequences length to be included in the analyzed table is often used in codon usage studies. The first rationale for this is that some genuine very small coding sequences present in amino acid operon leaders have an usual amino acid composition. For instance, the *pheL* gene coding for the phenylalanine operon leader peptide in *Escherichia coli* is extremely rich in Phe:

```
if(final.query3){
  choosebank("emglib")
  pheL <- query("pheL", "N=ECOLICG.PHEL")
  pheL.aa <- getTrans(pheL$req[[1]])
  save(pheL.aa, file = "local/pheL.rda")
  closebank()
} else {
  load("local/pheL.rda")
}
cat(aaa(pheL.aa))
```

```
Met Lys His Ile Pro Phe Phe Phe Ala Phe Phe Phe Thr Phe Pro Stp
```

A second rationale is that small coding sequences could be ELFs, an acronym for Evil Little F...ellows [OCH 02], that is, coding sequences predicted by annotation algorithms but which are not genuine coding sequences. A threshold of 100 codons is sometimes used, and, as a matter of comparison, the expected average reading-frame lengths in random sequences are about 13 codons and 67 codons for 25% and 75% GC content, respectively [OLI 96]. Here, we will not use a static threshold, but will instead use a varying threshold in section 2.6.

From a codon point of view, it could be tempting too to remove those which are poorly documented. For instance in the present data set (see Figure 2.3), there are only 119 codons cgg for Arg, that is much less than the 850 coding sequences in the data set. Again, we will not use a static threshold but a varying threshold in section 2.6.

Stop codons are often removed in codon usage studies. For synonymous codon usage I see no good reason to do so since there are interesting things to note. For instance Figure 2.3 shows that taa is more frequent than tag and tga, as expected in a GC-poor genome. The question is more disputable for amino acid usage since stop codons are not translated. I found it useful to keep them however, since it includes the pseudo-amino acid Stp. Since there is by definition only one Stp by protein, its relative frequency is inversely proportional to protein size. Then, if there is a protein size related factor for amino acid variability, the pseudo Stp amino acid may help to detect it. The stop codons were not removed here.

Lastly, choosing to remove or keep the initial atg codon coding for the N-terminal formyl methionine is rather a question of taste since the rules that govern its removal *in vivo* are not documented for most bacterial species. It was not removed here.

2.2. Running global correspondence analysis

We run the analysis with the function dudi.coa() from package ade4[2] and choose to keep five factors:

```
library(ade4)
tuco.coa <- dudi.coa(df = tuco, scannf = FALSE, nf = 5)
```

---

2 Correspondence analysis is also available with corresp() from MASS [VEN 02], ca() from ca [NEN 07], CA() from FactoMineR [LÊ 08] and afc() from amap [LUC 14].

Note that in the interactive mode the call could be tuco.coa <- dudi.coa(tuco) in which case the eigenvalue scree plot is shown to the user so they may choose the number of factors to keep. In this book, the call is in a non-interactive mode such that we must add scannf = FALSE to neutralize the eigenvalue scree plot display and nf = 5 to keep 5 factors. In all cases, the first argument tuco is an object from class data.frame[3] which entry tuco[i, j] is the count n_{ij} in Table 1.1. The result is a named list of 12 elements:

```
names(tuco.coa)
```

```
 [1] "tab"   "cw"    "lw"    "eig"   "rank" "nf"    "c1"    "li"    "co"    "l1"
         "call"
[12] "N"
```

The meaning of these elements is explained in the documentation of the function dudi.coa(), which is available in your ℝ's prompt via ?dudi.coa. Those used in this book are as follows:

– eig: For *eigenvalues*, this is the vector of eigenvalues λ_i associated with factor F_i, which is very important since eigenvalues are proportional to the total inertia factors account for.

– li: For *lines*, this is a table containing the coordinates of lines on kept factors, for instance li[, 1] for the coordinates on the first factor F_1.

– lw: For *line weights*, a vector of marginal totals divided by the grand total, $n_{i\bullet}/n_{\bullet\bullet}$, with the notation from Table 1.1.

– co: For *columns*, this is a table containing the coordinates of columns on kept factors, for instance co[, 2] for the coordinates on the second factor F_2.

– cw: For *column weights*, a vector of marginal totals divided by the grand total, $n_{\bullet j}/n_{\bullet\bullet}$, with the notation from Table 1.1.

---

3 It could in fact be any class, such as matrix, that is easily coerced into a data.frame.

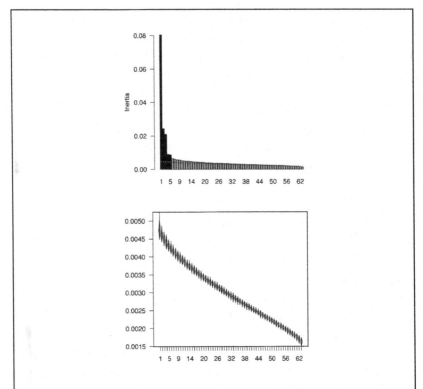

Figure 2.4. *Global CA scree plot. For a color version of this figure, see www.iste.co.uk/lobry/multivariate.zip*

NOTES ON FIGURE 2.4.– The eigenvalue scree plot for CA of the table of codon usage is given in the top panel. The first five bars along the axes are in black because five factors were kept in the analysis. The red points are the mean values under the independency hypothesis obtained from 1,000 simulations. The distributions of the expected values are represented with violin plots [HIN 98] in the bottom panel with a zoomed y-scale. ℝ code available in Appendix 1, section A1.3.3.

The eigenvalue scree plot is shown in Figure 2.4. We compute the percentage and the cumulated percentage of inertia taken into account by the first five factors:

```
(pti <-100*tuco.coa$eig[1:5]/sum(tuco.coa$eig))
```

```
[1] 23.909870 7.231221 6.265265 2.683554 2.585225
```

```
cumsum(pti)
```

```
[1] 23.90987 31.14109 37.40636 40.08991 42.67513
```

The first five factors take into account 23.91%, 7.23%, 6.27%, 2.68% and 2.59% of total inertia, respectively. Overall, with only five factors, we have extracted 42.7% of initial variability. The percentage of total variability taken into account by a factor is usually given in the axis legend of the factor in factorial maps, and we will stick here to this convention with the following utility function:

```
mklab <- function(x, n){
    paste("F", n, ": ", signif(100*x$eig[n]/sum(x$eig), 3), "%", sep = "")
}
mklab(tuco.coa, 1)
```

```
[1] "F1: 23.9%"
```

Note, however, that an alternative method is possible. The expected mean value from a random variable distributed according to a χ^2 distribution with k degrees of freedom is k, so that we can compute the expected total variability under the null hypothesis 1.1 as:

```
((nrow(tuco) - 1)*(ncol(tuco) - 1)/sum(tuco) -> exptoti)
```

```
[1] 0.1882536
```

A different way of appreciating the contribution of factors is then to remove this expected inertia under H_0 from the total inertia:

```
(pti2 <-100*tuco.coa$eig[1:5]/(sum(tuco.coa$eig) - exptoti))
```

```
[1] 54.313009 16.426245 14.232006  6.095888  5.872526
```

```
cumsum(pti2)
```

```
[1] 54.31301 70.73925 84.97126 91.06715 96.93967
```

This corresponds graphically to removing the area under the red dots in the top panel of Figure 2.4. We may say here, for instance, that F_1 takes into account 54.3% of the *structured* variability. This approach is not standard but could be useful as a safeguard: if the cumulated structured variability of the remaining factors is over 100%, then it means that we are digging too far and that less factors should be considered. We shall see an example of such a situation in section 2.6.2.

2.3. The missing factor F_0

Long before the genomic era[4], it was already known that the GC content of bacterial genomes is a major factor in between-species variability [BEL 58, SUE 62], strong enough to modulate the amino acid content of proteins [SUE 61, LOB 97]. *B. burgdorferi* is an extremely low GC bacteria, so why is this salient feature not the first factor returned by CA? If the obvious is not returned by CA, "I want my money back".

The table of codon counts is analyzed by CA for its departure from the null hypothesis 1.1. The expected codon counts n_{ij}^t in equation 1.3 are based on the marginal row and column totals that are already GC poor. With a single species under study, CA can only extract factors of within-species

---

4 From my point of view, the complete genome of *Rickettsia prowazekii* in 1998 is the real start of the genomic era because for the first time a scientific question, the origin of mitochondria, was present in the title of the publication [AND 98].

variability. Without any external reference, it is impossible to appreciate the between-species variability. If more bacterial species are present in the analyzed table (e.g. the 457 species in [LOB 06]), the GC content will of course be returned as the major factor of between-species variability.

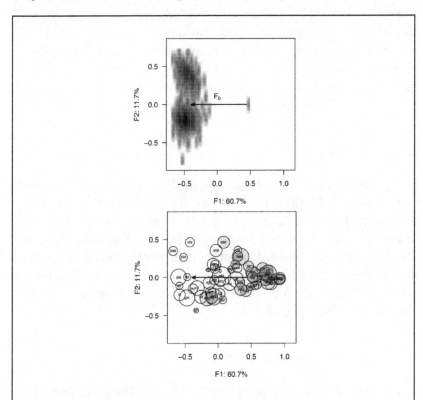

Figure 2.5. *Do not throw the baby out with the bathwater*

NOTES ON FIGURE 2.5.– The "baby" in this figure is the internal structure of the data set visible in the top panel with the two groups of coding sequences. The bathwater is the missing factor F_0 depicted by the arrow. These figures were obtained by adding to the original data set an artificial coding sequence with the same number of codons as in the original data set but with a purely uniform usage of them (it therefore has a GC content of 50%). The bottom panel, in codon space, with GC-ending codons in gray and AT-ending codons in white, shows that *B. burgdorferi* is an extremely GC-poor bacteria. ⓡ code available in Appendix 1, section A1.3.4.

Put in more general terms, if all factors in the analyzed contingency table are subject to the same trend, this trend will not appear as a factor in CA. It is therefore always advisable to carefully read the raw data as in Figure 2.3. Figure 2.5 shows that the missing factor F_0 is easy to represent by adding an extra row to the original data set.

2.4. First factor

The first factor is so dominant in the eigenvalue scree plot (see Figure 2.4) that it deserves its own analysis. This is somewhat frustrating, as multivariate analyses are expected to produce nice factorial maps and that none is possible with a single factor at hand. However, it is worth mentioning that it is perfectly possible to study a *single* factor at once. This is especially useful for data sets where a single factor overpowers the scree plot.

2.4.1. *Coding sequence point of view*

As shown by the top panel in Figure 2.6, the distribution of coding sequences on the first factor of CA is clearly bimodal. What CA shows here is that the most important factor of variability for codon usage in *B. burgdorferi* is the consequence of two subpopulations of coding sequences using their own dialect. This is typically a structure undetectable from a global codon usage table (see Figure 2.3), but which we can reveal here by a multivariate analysis.

To characterize further the two groups, we use as an illustrative variable the orientation of the coding sequences with respect to replication in the bottom panel in Figure 2.6. The two subpopulations are therefore defined by their orientation with respect to replication: the most important factor of variability is the difference in codon usage between the genes in the leading group and in the lagging one.

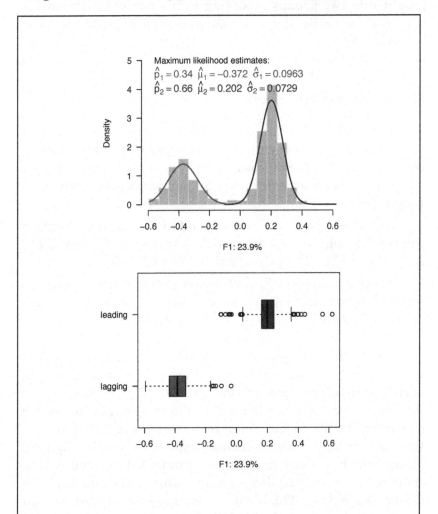

Figure 2.6. *Bimodal distribution on the first factor. For a color version of this figure, see www.iste.co.uk/lobry/multivariate.zip*

NOTES ON FIGURE 2.6.– The distribution of the 850 coding sequences on the first factor of CA is given in the top panel, with a maximum likelihood fit to a mixture of two normal distributions. The bottom panel is the same distribution when the coding sequences are split into two groups according to their orientation with respect to replication. ℝ code available in Appendix 1, section A1.3.5.

2.4.2. *Codon point of view*

Figure 2.7 shows the distribution of codons on the first factor. Because CA is symmetric with respect to the rows and columns, we can directly interpret this figure in conjunction with the coordinates for the coding sequences shown in Figure 2.6:

– negative values on F_1 correspond to the lagging group, so the top 10 codons they are enriched in are as follows: CTA, AAC, CTC, ATC, TAC, CAC, ACA, ATA, TTC, TGC;

– positive values on F_1 correspond to the leading group, so the top 10 codons they are enriched in are as follows: CGT, GTT, GGT, AGG, TAG, CGG, GAG, AAG, TGT, CAG.

We note from the top 10 codons that the lagging group is enriched in AC-ending codons while the leading group is enriched in GT-ending codons, which we may use as an illustrative variable in Figure 2.7. The most important factor of variability for codon usage in *B. burgdorferi* is therefore the opposition between coding sequences in the lagging group, which are enriched in AC-ending codons, and coding sequences in the leading group, which are enriched in GT-ending codons. This factor is strong enough that we can almost always correctly predict the orientation of a coding sequence with respect to replication just from its codon usage.

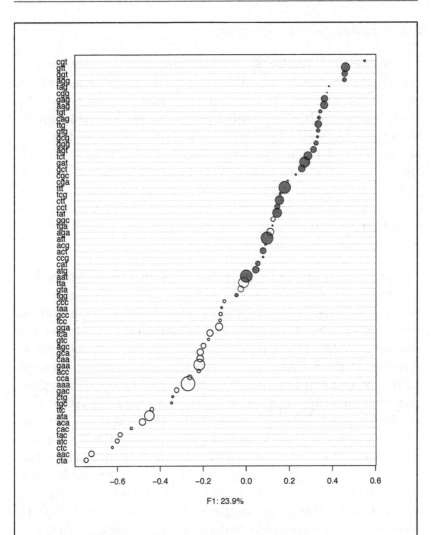

Figure 2.7. *GT- versus AT-ending codons*

NOTES ON FIGURE 2.7.– Distribution of codons on the
first factor of CA with GT-ending codons in gray and
AC-ending codons in white. The area of the circles is
proportional to the total number of codons in the data set.
ℝ code available in Appendix 1, section A1.3.6.

2.4.3. *Biological interpretation*

Because substitutions in the third codon position are mainly synonymous, this suggests that F_1 is the result of an asymmetric directional mutation pressure between the two DNA stands [LOB 96a], which is a common phenomenon in bacteria [LOB 02], but seldom strong enough to be the first factor of codon usage variability as it is here in *B. burgdorferi*. We also note that the leading group is more populated than the lagging group, also a common feature in bacteria [MAO 12], which is not the consequence of a mutation pressure but of a selective one that we will discuss with the analysis of the second factor of variability.

Out of many explanations [FRA 99] that were published to explain the observed compositional biases between the leading and the lagging strand, the cytosine deamination theory is now well supported by experimental evolutionary studies [BHA 16]. Introduced first for mitochondrial genomes [BRO 82], the cytosine deamination theory is based on the experimental evidence that cytosine reacts with water, creating uracil, and the rate of this reaction in single-stranded DNA is more than 100 times the rate in double-stranded DNA [LIN 74, FRE 90, SHE 94]. During replication, the template lagging strand is protected by the newly synthesized leading strand while the template leading strand has to afford a transient single strand state waiting for the newly synthesized lagging strand to be long enough to recover a double-stranded state [NOS 83, BAK 92, MAR 92]. The net result is an excess of C $\rightarrow$ T mutations in the leading strand, resulting in an excess of G over C and T over A.

2.5. Second and third factors

Because the eigenvalues associated with F_1 and F_2 are very close in the scree plot shown in Figure 2.4, we choose here to represent data in the F_2–F_3 plane. This may help to detect a degenerate situation such as the one depicted in Figure 1.5.

2.5.1. *Coding sequence point of view*

The second factorial map in Figure 2.8 shows that there is a sub-population of coding sequences on the top of F_3 and shifted to the right on F_2, and representing about 12% of the total number of coding sequences (see Figure 2.9), which is close to the proportion of integral membrane protein usually found in bacteria[5]. A simple way to check that this subpopulation corresponds to integral membrane proteins is to compute the hydropathy index from KYTE and DOOLITTLE [KYT 82], also known as the GRAVY score. This is a linear form, s, on codon frequencies:

$$s = \sum_{i=1}^{64} \alpha_i f_i$$

where α_i is the coefficient for the amino acid encoded by codon i and f_i the relative frequency of codon number i. The values for α_i are given by the element KD from data EXP. Then, because of ℝ's built-in matrix operators, computing this linear form is compact and straightforward:

```
data(EXP)
kd <- (tuco/rowSums(tuco)) %*% EXP$KD
head(kd)


                    [,1]
BORBUCG.BB0001   0.40209424
BORBUCG.BB0002   0.12186589
BORBUCG.BB0003  -0.08483516
BORBUCG.BB0004  -0.33946488
BORBUCG.BB0005  -0.26355932
BORBUCG.BB0006   0.89680000
```

---

5 For example, 11% for one species in [LOB 94] and 13.3% for 25 species in [LOB 03].

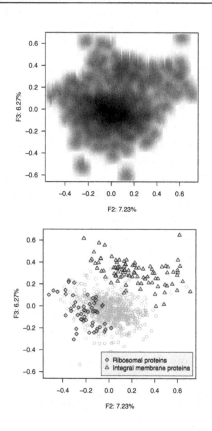

Figure 2.8. *Second factorial map for global CA. For a color version of this figure, see www.iste.co.uk/lobry/multivariate.zip*

NOTES ON FIGURE 2.8.– Location of the 850 coding sequences on the second factorial map of global CA. The top panel uses an estimate of the local density of coding sequences coordinates. In the bottom panel, the two illustratives variables are the sequences coding for ribosomal proteins and the sequences coding for integral membrane proteins. ℝ code available in Appendix 1, section A1.3.7.

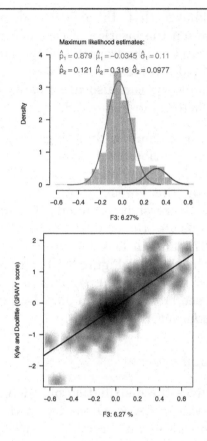

Figure 2.9. *Third factor of global CA. For a color version of this figure, see www.iste.co.uk/lobry/multivariate.zip*

NOTES ON FIGURE 2.9.– Top panel: Distribution of the 850 coding sequences on the third factor of CA and a maximum likelihood fit with a mixture of two normal distributions. Bottom panel: Correlation between KYTE and DOOLITTLE hydropathy index [KYT 82] and the coordinates on the third factor of global CA analysis. ℝ code available in Appendix 1, section A1.3.8.

Figure 2.9 shows that there is a strong correlation ($r^2 = 0.659$) between the location of the coding sequences on the third factor and the hydropathy of the encoded protein. Using a threshold[6] value $s > 0.5$ to predict integral membrane proteins, we can also use this as an illustrative variable on the factorial in Figure 2.8.

As a proxy for a gene with a high expressivity level, sequences coding for ribosomal proteins are plotted as an illustrative variable in the bottom panel of Figure 2.8. They are all located on the lowest values on F_2 suggesting an underlying gene expressivity gradient. A simple and usual method in codon usage studies to check for this is to compute the CAI [SHA 87], and to look for the correlation with the location on F_2. The result in Figure 2.10 is however extremely disappointing ($r^2 = 4.85 \times 10^{-6}$). This is not the regular result; we shall see later in section 3.2 why the CAI does not work for *B. burgdordefi*.

2.5.2. *Codon point of view*

Figure 2.11 gives the distribution of codons on the second factor. Again, because CA is symmetric with respect to the rows and columns, we can directly interpret this figure in conjunction with the coordinates for the coding sequences in Figure 2.8:

– negative values on F_2 correspond to sequences coding for ribosomal proteins, so the top 10 codons they are enriched in are as follows: **CGT, GCG, CGA, GCT, GGT, GCA, CAG, AGA, GGA, CGC**;

– positive values on F_2 correspond to sequences not coding for ribosomal proteins, so the top 10 codons they are enriched in are as follows: TTT, **TGA**, TTC, TTG, TAT, TTA, AAT, TAC, TCC, AGT.

---

6 This threshold value is based on inspection of Figure 8 in [KYT 82].

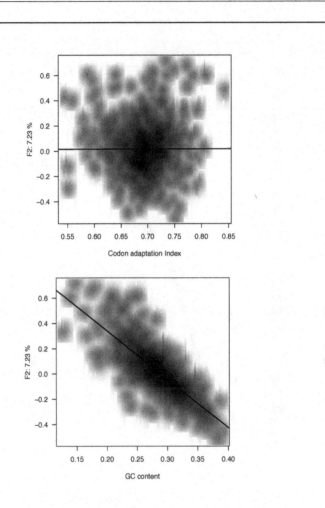

Figure 2.10. *Second factor of global CA*

NOTES ON FIGURE 2.10.– Correlation for the 850 coding sequences between the location on the second factor of CA. (i) Top panel: the value of the codon adaptation index [SHA 87]; (ii) bottom panel: the GC content of the coding sequence. ℝ code available in Appendix 1, section A1.3.9.

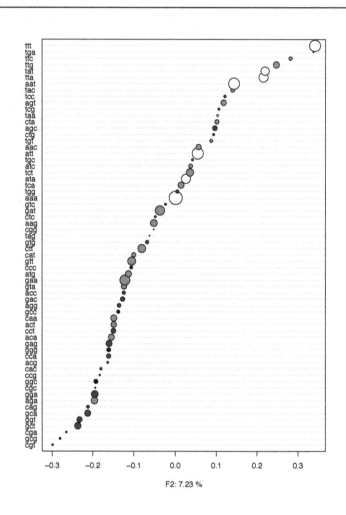

Figure 2.11. *Second factor of global CA*

NOTES ON FIGURE 2.11.– Distribution of codons on the second factor of CA with a darker gray for GC-rich codons. The area of the circles is proportional to the total number of codons in the data set. ℝ code available in Appendix 1, section A1.3.10.

Noting from the top 10 codons that the ribosomal group is enriched in **GC**-rich codons, we use this as an illustrative variable in Figure 2.11. The second most important factor of variability for codon usage in *B. burgdorferi* is therefore a gradient of GC content. Turning back to the coding sequence point of view, we can see that in the bottom panel of Figure 2.10 there is a strong correlation ($r^2 = 0.652$) between the location on F_2 and the GC content of the coding sequences.

We already know from the sequence point of view that F_3 discriminates integral membrane proteins, so we use the physicochemical properties of the encoded amino acid as an illustrative variable for the codon location on F_3 in Figure 2.12. Integral membrane proteins are enriched in non-polar amino acids and avoid charged amino acids.

2.5.3. *Biological interpretation*

The third factor F_3 is the result of a selective pressure to maintain the sub-cellular location of proteins yielding a hydropathy gradient in the amino acid composition of the encoded proteins from the location of the hydrophilic cytoplasm to the hydrophobic membrane.

The second factor F_2 could be a gradient of gene expressivity with highly expressed genes enriched in GC-rich codons. There are, however, no data for tRNA intracellular content in *B. burgdorferi* to check for this, so henceforth I will refer to the "gene expressivity" factor within quotes.

Turning back to the first factor F_1, we want to discuss the excess of genes in the leading strand. Many factors have been put forward to explain this excess [ZHE 15], the oldest being the result of a selective pressure to avoid head-on collisions between the RNA polymerase and the replication fork [BRE 88, ZEI 90, LIU 95, MER 12, GAR 16, HAM 16].

Under this hypothesis, the selective pressure for a gene to be in the lead should increase with expressivity, and early reports from the first complete bacterial genomes supported this [MCL 98, PER 98]. The proportion of genes in the leading strand is also expected to increase with the specific growth rate, which was the general trend observed in a set of 104 bacteria [MAO 12]. There are, however, many confounding factors that may interact with gene expressivity and complicate the interpretation (see references 4–15 in [GAO 17]). We want to test here the potential link between gene expressivity and gene orientation bias, first playing with our subset of genes coding for ribosomal proteins.

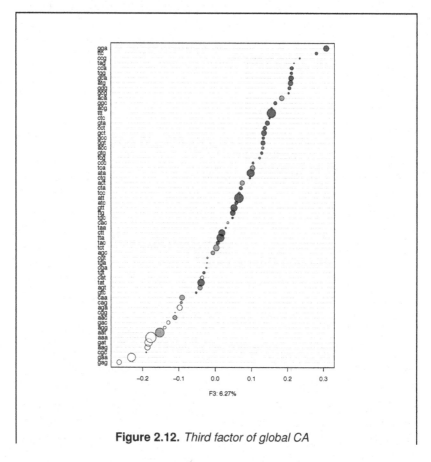

Figure 2.12. *Third factor of global CA*

NOTES ON FIGURE 2.12.– Distribution of codons on the third factor of CA with codons coding for non-polar amino acids in black and codons coding for charged amino acids in white. The area of the circles is proportional to the total number of codons in the data set. ℝ code available in Appendix 1, section A1.3.11.

```
ribnames <- readLines("local/ribnames.txt")
isrib <- rownames(tuco) %in% ribnames
(riblea <- table(list(isrib, leading), dnn = c("isrib", "isleading")))
```

```
       isleading
isrib   FALSE TRUE
  FALSE   284  513
  TRUE      2   51
```

```
chisq.test(riblea)
```

```
        Pearson's Chi-squared test with Yates' continuity correction
data:  riblea
X-squared = 21.19, df = 1, p-value = 4.16e-06
```

The hypothesis of independence between coding for a ribosomal protein and its position in the leading strand is rejected here, and as shown by Figure 2.13 the excess of genes in the leading strand is enhanced in the ribosomal group, as expected.

2.6. Fourth and fifth factors

We are digging now in the far-east of the eigenvalue scree plot in Figure 2.4. We want to check if these factors are not due to a rare codon or a small coding sequence effect, yielding an acute correspondence as the rare Cys amino acid that was found in only one protein in the example used in Chapter 1. A simple way to study this is to "torture" the data set by progressively removing rows or columns starting from those with the smallest marginals and running CA again and again. If the fourth and fifth factors are due to an acute correspondence, their contribution to structured inertia will vanish.

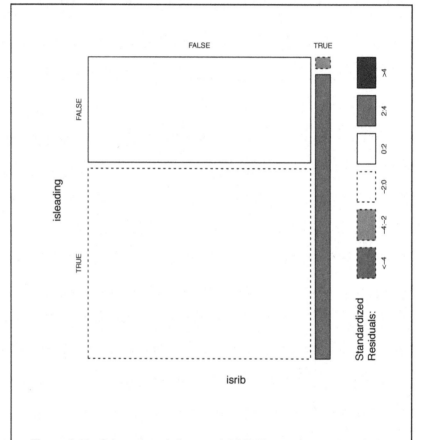

Figure 2.13. *Orientation of ribosomal CDS. For a color version of this figure, see www.iste.co.uk/lobry/multivariate.zip*

NOTES ON FIGURE 2.13.– This is a mosaicplot [HAR 84, EME 98, FRI 94] representation of the contingency table crossing CDS orientation (leading vs. lagging) with the members of the ribosomal subset defined in Table 2.1. ⓡ code available in Appendix 1, section A1.3.12.

2.6.1. *Coding sequence point of view*

Figure 2.14 shows that the structured inertia is very stable even when deleting all sequences less than 600 codons, such that only 10% of initial sequences representing 25% of initial codons are kept. The only trend is the decrease of the contribution of F_3, the hydropathy gradient factor. Integral membrane proteins are on average smaller than the cytoplasmic ones so that they are preferentially removed, decreasing the contribution of F_3 to structured inertia. Figure 2.14 shows that over 500 amino acids the proportion of integral membrane proteins begins to vanish. It is likely that there is bias a selective against integral membrane proteins being too large so that they stay in the hydrophobic phospholipid bilayer. This is interesting but a more direct representation than the abstract evolution of the eigenvalues in a scree plot could be useful here. I postpone[7] this discussion to section 4.3.2.1.

2.6.2. *Codon point of view*

The bottom panel in Figure 2.14 shows that the contribution of the first five factors to structured inertia quickly exceeds 100% when deleting codons. Since we are starting with codons with the smallest marginal totals, it means that we are seeing a rare codon effect. This kind of effect is unlikely to be interesting: we are looking for factors that modulate the usage of many codons at once, not just a few of them. We are too greedy here by keeping five factors, and only three should be considered.

---

7 Sorry for this, but I need some cliffhangers for my book to yield a blockbuster in Hollywood. Spoiler alert: BOX and COX [BOX 64] are the villains here.

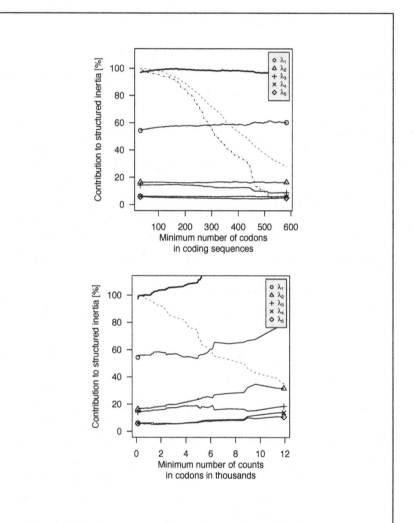

Figure 2.14. *Data torture. For a color version of this figure, see www.iste.co.uk/lobry/multivariate.zip*

NOTES ON FIGURE 2.14.– These graphics show the stability of the eigenvalue scree plot when the data set is reduced. In the top panel, the coding sequences are progressively removed from the data set, starting from the smallest to the largest and stopping when only 10% are left. The dotted lines starting from 100% are on top the fraction of remaining codons (in red) and just below the fraction of remaining codons coding for integral membrane proteins (in blue). The heavy line is the contribution of the first five factors to structured inertia. In the bottom panel, the codons are progressively removed from the data set, starting from the less frequent to the more frequent and again stopping when only 10% are left. The dotted line is the fraction of remaining codons. The heavy line that dangerously exceeds 100% is the contribution of the first five factors to structured inertia. ® code available in Appendix 1, section A1.3.13.

Within and Between Correspondence Analysis

3.1. Running the analyses

Within-block correspondence analysis (WCA) [BEN 83] is "the method of choice" [EME 10] for synonymous codon usages studies as demonstrated by a methodological comparison on 241 bacterial genomes [SUZ 08]. The use of WCA in genomic studies is relatively recent [LOB 03, CHA 05]. As compared to the global codon usage analysis of previous chapter, WCA focuses on the within-amino acid variability, that is, the synonymous variability. WCA is complemented by between-block correspondence analysis (BCA), which focuses on the between-amino acid variability, that is, the non-synonymous variability. Formally, BCA is the same as CA on amino acid usage. The factor defining the block structure, illustrated by Figure I.3, is defined with respect to the rows of the table in the ade4 package, so that our first task is to transpose the table before running the analyses:

```
ttuco <- t(tuco) # Transpose the table of codon usage
ttuco.coa <- dudi.coa(ttuco, scan = FALSE, nf = 5)
aaname <- sapply(rownames(ttuco), function(x) translate(s2c(x)))
facaa <- as.factor(aaa(aaname))
```

```
ttuco.wca <- wca(ttuco.coa, facaa, scan = FALSE, nf = 1)
ttuco.bca <- bca(ttuco.coa, facaa, scan = FALSE, nf = 5)
```

The functions `wca()` and `bca()` correspond to WCA and BCA, respectively, as we may have expected from their names. Their first argument `ttuco.coa` is the result of CA on the transposed table of codon usage `ttuco`. Their second argument `facaa` is a qualitative variable that defines the row-block structure, defined here by applying the genetic code on codons:

```
rownames(ttuco)
```

```
[1]  "aaa" "aac" "aag" "aat" "aca" "acc" "acg" "act" "aga" "agc" "agg" "agt" "ata"
[14] "atc" "atg" "att" "caa" "cac" "cag" "cat" "cca" "ccc" "ccg" "cct" "cga" "cgc"
[27] "cgg" "cgt" "cta" "ctc" "ctg" "ctt" "gaa" "gac" "gag" "gat" "gca" "gcc" "gcg"
[40] "gct" "gga" "ggc" "ggg" "ggt" "gta" "gtc" "gtg" "gtt" "taa" "tac" "tag" "tat"
[53] "tca" "tcc" "tcg" "tct" "tga" "tgc" "tgg" "tgt" "tta" "ttc" "ttg" "ttt"
```

```
facaa
```

```
[1]  Lys Asn Lys Asn Thr Thr Thr Thr Arg Ser Arg Ser Ile Ile Met Ile Gln His Gln His
[21] Pro Pro Pro Pro Arg Arg Arg Arg Leu Leu Leu Leu Glu Asp Glu Asp Ala Ala Ala Ala
[41] Gly Gly Gly Gly Val Val Val Val Stp Tyr Stp Tyr Ser Ser Ser Ser Stp Cys Trp Cys
[61] Leu Phe Leu Phe
21 Levels: Ala Arg Asn Asp Cys Gln Glu Gly His Ile Leu Lys Met Phe Pro Ser ... Val
```

Inspection of Figure 3.1 shows that there is only one factor left at the synonymous codon usage level while there are still five left at the amino acid usage one. We also note that there is more variability taken into account at the synonymous level than at the amino acid level:

```
#all.equal(sum(ttuco.coa$eig), sum(ttuco.wca$eig) + sum(ttuco.bca$eig))
        is TRUE
100*sum(ttuco.wca$eig)/sum(ttuco.coa$eig)
```

```
[1] 59.78634
```

```
100*sum(ttuco.bca$eig)/sum(ttuco.coa$eig)
```

```
[1] 40.21366
```

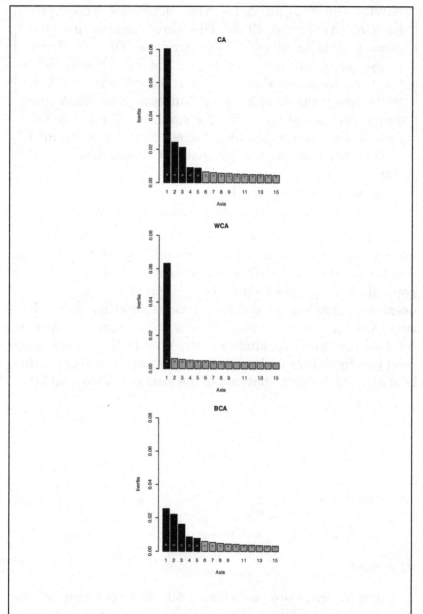

Figure 3.1. *Synonymous codon and amino acid usage. For a color version of this figure, see www.iste.co.uk/lobry/multivariate.zip*

NOTES ON FIGURE 3.1.– The eigenvalue scree plot for CA, WCA and BCA. The three figures use the same y-scale to allow for comparisons. The top figure is the same as shown in Figure 2.4 except that only the first 15 eigenvalues are represented here. WCA is the synonymous codon usage analysis and BCA the amino acid usage analysis. The small red points are the mean values under the independency hypothesis obtained from 1,000 simulations for global CA and from 1,000 simulations for WCA and BCA. ℝ code available in Appendix 1, section A1.4.1.

We have therefore about 60% of the variability at the synonymous level and the remaining 40% at the non-synonymous level. This is however not a completely fair point of view because we have $(850 - 1) \times (64 - 21) = 849 \times 43$ degrees of freedom at the synonymous level as opposed to only $(850 - 1) \times (21 - 1) = 849 \times 20$ at the non-synonymous level, so that more variability is expected at the synonymous level just by chance under H_0. A simple way to correct for this is to express the contributions to the structured variability:

```
(nrow(tuco) - 1)*(ncol(tuco) - 1)/sum(tuco) -> exptoti
(nrow(tuco) - 1)*(ncol(tuco) - length(levels(facaa)))/sum(tuco) ->
     exptotiw
(nrow(tuco) - 1)*(length(levels(facaa)) - 1)/sum(tuco) -> exptotib
# all.equal(exptoti, exptotiw + exptotib) is TRUE
100*(sum(ttuco.wca$eig) - exptotiw)/(sum(ttuco.coa$eig) - exptoti)
```

[1] 49.01913

```
100*(sum(ttuco.bca$eig) - exptotib)/(sum(ttuco.coa$eig) - exptoti)
```

[1] 50.98087

There is therefore an almost 50–50 repartition of the structured variability between the synonymous and non-synonymous level.

3.2. Synonymous codon usage (WCA)

3.2.1. *The first and unique factor* F_1

3.2.1.1. *Coding sequences point of view*

The distribution of coding sequences on the first factor is bimodal with the two groups corresponding with almost no overlap to the leading and the lagging orientation with respect to replication (see Figure 3.2).

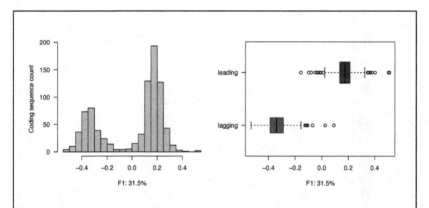

Figure 3.2. *First factor for synonymous codon usage. For a color version of this figure, see www.iste.co.uk/lobry/multivariate.zip*

NOTES ON FIGURE 3.2.– Distribution of the 850 coding sequences coordinates on the first factor of synonymous codon analysis (WCA). The bimodal distribution on the left is well explained on the right, by the orientation of genes with respect to replication. ℝ code available in Appendix 1, section A1.4.2.

3.2.1.2. *Codon point of view*

Figure 3.3 shows that the leading group is enriched in GT-ending codons while the lagging one is enriched in AC-ending codons.

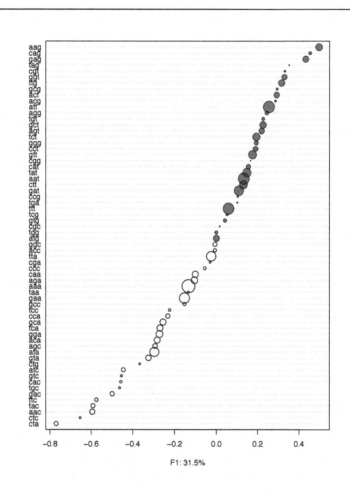

F1: 31.5%

Figure 3.3. *GT-ending versus AT-ending codons*

NOTES ON FIGURE 3.3.– Distribution of codons on the first factor of WCA with GT-ending codons in gray and AC-ending codons in white. The area of the circles is proportional to the total number of codons in the data set. ℝ code available in Appendix 1, section A1.4.3.

3.2.1.3. *Biological interpretation*

There is a very strong asymmetric directional mutation pressure within this genome. This is unusual because, in unicellular organisms, the most important factor of synonymous codon usage variability is usually linked to gene expressivity: frequent codons correspond to tRNA with a high intracellular concentration and this trend is exacerbated for highly expressed genes [IKE 81, GOU 82, HOL 86, SHA 86, AND 90, KAN 99]. For instance, out of 241 bacterial genomes [SUZ 08], the first axis of WCA was found to be linked to an asymmetric directional mutation pressure in only 38 cases (16%) and was more common as a second or third factor in 70 cases (29%). In Archaea, out of 67 genomes [EME 10], the first factor of WCA was found to be linked to an asymmetric directional mutation pressure in only five cases (8%). *B. burgdorferi* is then exceptional because there is only one factor of synonymous codon variability and this factor is due to a strand-specific mutation pressure. *B. burgdorferi* is a genomic monster.

We now understand the surprising result from section 2.5.1 in which the codon adaptation index (CAI) [SHA 87] was poorly correlated with the "gene expressivity" factor from global CA. The CAI is defined as

$$\ln(\text{CAI}) = \sum_{i=1}^{59} f_i \ln w_i, \qquad [3.1]$$

where f_i is the relative frequency of codon of kind i in the coding sequences[1], and w_i is the ratio of the frequency of codon of kind i to the frequency of the major codon for the same amino acid. In others words, CAI is based on

---

1 Three stop codons and the single codon tgg for Trp and atg for Met are not taken into account, such that there is a total of 59 terms in equation [3.1].

synonymous codon usage only, and since there is no "gene expressivity" factor at the synonymous level in *B. burgdorferi*, there is no way for the CAI to reflect a factor that is only present at the *non-synonymous* level. *B. burgdorferi* is a genomic monster because in bacteria the "gene expressivity" factor is usually present at the synonymous level and, to a lesser extent, at the non-synonymous level [SHP 89, LOB 94]. The symmetric and asymmetric directional mutation pressures are so strong in this genome that there is no more room left for others factors at the synonymous level.

3.3. Amino acid usage (BCA)

3.3.1. *The missing factor F_0*

Figure 3.4 shows that the missing factor F_0 for amino acid usage is, again, the genomic GC content, as for the global codon usage analysis in Figure 2.5. *B. burgdorferi* is a GC-poor bacteria and as such avoids amino acids encoded by GC-rich codons. This is a well-known factor in bacteria that was documented early [SUE 61]. It does not appear as a factor in BCA because all proteins are subject to the same directional mutation pressure.

3.3.2. *The ugly* (F_1, F_2, F_3) **ménage à trois**

Previous studies have suggested that the most important factor for amino acid usage variability within the *B. burgdorferi* genome is linked to an asymmetric directional mutation pressure [ROC 99b, LAF 99, MAC 99b]. This is not so simple because we have a rather unpleasant highly degenerated situation here, illustrated by Figure 3.5.

3.3.2.1. *Coding sequences point of view*

The left panels in Figure 3.5 are the factorial maps in coding sequence space. Instead of the three axes F_1, F_2 and F_3

returned by BCA, we would like to have something like $G_1 = -F_1 - F_2$ for the "gene expressivity" gradient, $G2 = -F_1 + F_3$ for the hydropathy gradient and $G_3 = F_1 + F_3$ to discriminate between the leading and lagging group.

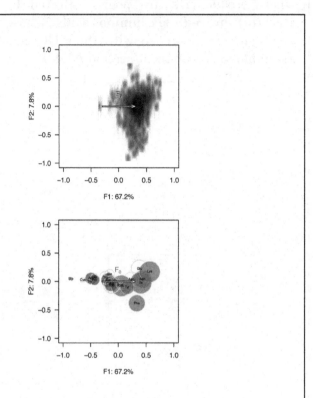

Figure 3.4. *The missing factor for amino acid usage. For a color version of this figure, see www.iste.co.uk/lobry/multivariate.zip*

NOTES ON FIGURE 3.4.– These figures were obtained by adding to the original data set an artificial coding sequence with the same number of codons as in the original data set but with a pure uniform usage of them (it therefore has a GC content of 50%). The top panel shows that the average amino acid composition is different from what would be expected for a uniform codon usage sequence. In the bottom panel, the amino acids were classified into three classes according to [LOB 97]: (i) in blue, four amino acids whose frequencies monotonously increase with GC-content, (ii) in white, 10 amino acids whose frequencies are poorly affected by GC-content and (iii) in red, six amino acids whose frequencies monotonously decrease with their GC content. ℝ code available in Appendix 1, section A1.4.4.

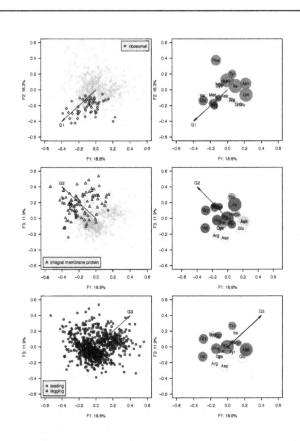

Figure 3.5. Ménage à trois. *For a color version of this figure, see www.iste.co.uk/lobry/multivariate.zip*

NOTES ON FIGURE 3.5.– The three factors G_1, G_2 and G_3 we would like to have instead of those returned by BCA for amino acid usage. See text section 3.3.2 for comments. ℝ code available in Appendix 1, section A1.4.5.

3.3.2.2. *Aminoacid point of view*

Because CA is symmetric, the degeneracy present in sequence space is also present in the amino acid space, as illustrated by the right panels in Figure 3.5.

3.3.2.3. *Methodological point of view*

The example of degeneracy in Figure 1.5 may look artificial, but the present data set of amino acid usage in *B. burgdorferi* shows that degeneracy does happen with actual data, and moreover this is a degeneracy in $\mathbb{R}^3$. I am unsure of the odds of such a situation, but again *B. burgdorferi* is a genomic monster. This is especially irritating because in global CA (see Figure 2.8), F_2 and F_3 were nicely represented. The drawback of BCA is that the contribution of F_1 to total inertia at the amino acid level is now close to the contribution of F_2 and F_3, yielding this degenerate structure[2].

---

2 There are many factor rotation methods such as the popular varimax rotation [KAI 58] that may help in degenerate situations, but their adaptation to CA is tricky [VAN 05] and I am not aware of an ℝ package implementing them.

Internal Correspondence Analysis

4.1. Running the analyses

Internal correspondence analysis (ICA) [CAZ 88, BÉC 05] deals with contingency tables that have a double partition on the columns and rows. In addition to the column-block structure defined by the amino acid in Chapter 3, we want to add a row-block structure to make a partition of the coding sequences according to their orientation with respect to replication and the leading and lagging groups[1]. We start with a WCA and BCA on the row-block structure, so there is no need to transpose data here:

```
tuco.coa <- dudi.coa(tuco, scannf = FALSE, nf = 5)
tuco.wca <- wca(tuco.coa, as.factor(leading), scannf = FALSE, nf = 4)
tuco.bca <- bca(tuco.coa, as.factor(leading), scannf = FALSE, nf = 1)
```

There is no difference here compared to section 3.1; the first argument of functions wca() and bca(), tuco.coa, is the result of CA on the table of codon usage tuco. Their second argument

---

1 Note that we may have more than two groups in the row-block structure, for instance the 25 species for the origin of the coding sequences [LOB 03] or the 18 tissues for tissue-specific coding sequences [SÉM 06].

as.factor(leading) is a qualitative variable that defines the row-block structure, corresponding here to the orientation of coding sequences with respect to replication.

```
summary(as.factor(leading))
```

```
FALSE   TRUE
 286    564
```

The eigenvalue scree plots are shown in Figure 4.1. About 77.4% of the global variability in codon usage is at the within-group level:

```
#all.equal(sum(tuco.coa$eig), sum(tuco.wca$eig) + sum(tuco.bca$eig))
       # is TRUE
100*sum(tuco.wca$eig)/sum(tuco.coa$eig)
```

```
[1] 77.36522
```

```
100*sum(tuco.bca$eig)/sum(tuco.coa$eig)
```

```
[1] 22.63478
```

This is, however, not a completely fair point of view because we have $63 \times 848 = 53{,}424$ degrees of freedom at the within-group level but only $63 \times 1 = 63$ at the between-group level, so more variability is expected at the within-group level just by chance under H_0. A simple way to correct for this is to express the contributions to the structured variability:

```
(nrow(tuco) - 1)*(ncol(tuco) - 1)/sum(tuco) -> exptoti
(nrow(tuco) - 1)*(ncol(tuco) - length(levels(as.factor(leading))))/
       sum(tuco) -> exptotiwg
(nrow(tuco) - 1)*(length(levels(as.factor(leading))) - 1)/sum(tuco) ->
       exptotibg
# all.equal(exptoti, exptotiwg + exptotibg) is TRUE
100*(sum(tuco.wca$eig) - exptotiwg)/(sum(tuco.coa$eig) - exptoti)
```

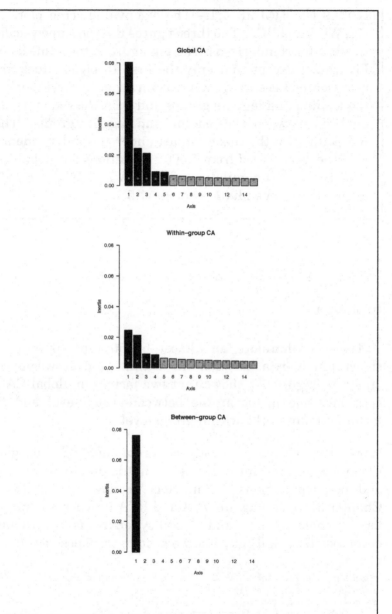

Figure 4.1. *Within- and between-group codon usage. For a color version of this figure, see www.iste.co.uk/lobry/multivariate.zip*

NOTES ON FIGURE 4.1.– The eigenvalue scree plot for CA, WCA and BCA. The three figures use the same y-scale to allow for comparisons. The top figure is the same as in Figure 2.4 except that only the first 15 eigenvalues are represented here. WCA is the codon usage analysis within the leading and lagging groups and BCA the codon usage analysis between the leading and lagging groups. The red points are the mean values under the independency hypothesis obtained from 1,000 simulations for global CA and from 1,000 simulations for WCA and BCA. ℝ code available in Appendix 1, section A1.5.1.

```
[1] 50.60182
```

```
100*(sum(tuco.bca$eig) - exptotibg)/(sum(tuco.coa$eig) - exptoti)
```

```
[1] 49.39818
```

There is therefore an almost 50–50 repartition of the structured inertia between the within- and between-group analyses. Figure 4.1 shows that five factors in global CA are split into one factor at the between-group level and four factors remain at the within-group level.

For the within- and between-group analyses for global codon usage, we want to now see what is due to synonymous and non-synonymous codon usage. Following the lines of Chapter 3, we run again WCA and BCA for each subanalysis for synonymous and non-synonymous variability decomposition, and this time we need to transpose data:

```
aaname <- sapply(colnames(tuco), function(x) translate(s2c(x)))
facaa <- as.factor(aaa(aaname))
facg <- as.factor(leading)
ttuco.coa <- dudi.coa(t(tuco), scan = FALSE, nf = 5)
ttuco.wca <- wca(ttuco.coa, facaa, scan = FALSE, nf = 1)
ttuco.bca <- bca(ttuco.coa, facaa, scan = FALSE, nf = 5)
```

```
ttuco.wca.wca <- wca(t(ttuco.wca), facg, scan = F, nf = 0)
ttuco.wca.bca <- bca(t(ttuco.wca), facg, scan = F, nf = 4)
ttuco.bca.wca <- wca(t(ttuco.bca), facg, scan = F, nf = 4)
ttuco.bca.bca <- bca(t(ttuco.bca), facg, scan = F, nf = 1)
```

The eigenvalue scree plots are shown in Figure 4.2 in terms of total inertia, and in Figure 4.2 in terms of structured inertia. Because there is an almost 50–50 repartition of the structured inertia between the synonymous and non-synonymous levels and between the within- and between-groups, there is no reason to start with the first or the latter. The choice here is to start with the synonymous and non-synonymous levels, so as to follow the structure of Chapter 3.

4.2. Synonymous codon usage

We already know from section 3.2 that there is only one factor for codon usage variability at the synonymous level. The within- and between-group analyses show that this factor is entirely due to the between-group effect: the central panel in Figure 4.2 shows that there is no more structure when the between-group effect is taken into account. Figure 4.4 shows that sequences in the leading strand are enriched in GT-ending codons, while sequences in the lagging strand are enriched in AC-ending codons.

4.3. Non-synonymous codon usage

4.3.1. *Between-group analysis*

Figure 4.5 shows that proteins encoded in the leading strand are enriched in GT-rich codons in the first and second codon positions, and those in the lagging strand have AC-rich codons in the first and second codon positions.

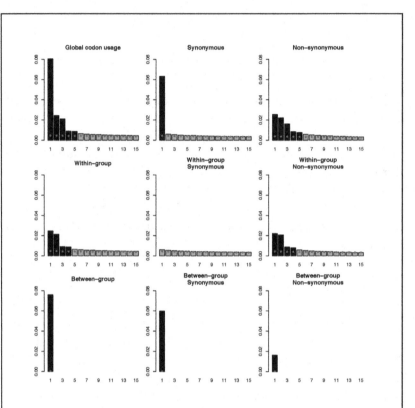

Figure 4.2. *Decomposition of codon usage factors. For a color version of this figure, see www.iste.co.uk/lobry/multivariate.zip*

NOTES ON FIGURE 4.2.– The eigenvalue scree plots for nine analyses associated with ICA. All figures use the same y-scale to allow for comparisons. The figures on the top row are the same as in Figure 3.1. The figures on the left column are the same as in Figure 4.1 but transposed. The small red points are the mean values under the independency hypothesis obtained by simulation. Kept factors are in black, and unkept factors are in gray. ℝ code available in Appendix 1, section A1.5.2.

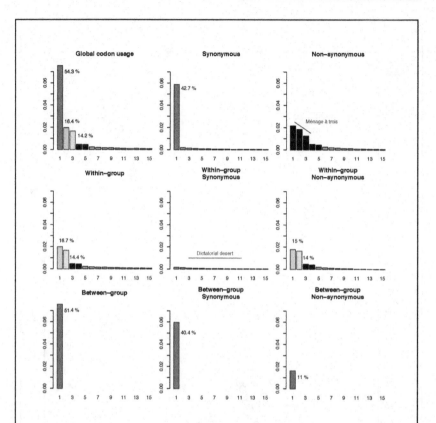

Figure 4.3. *Interpretation of codon usage factors. For a color version of this figure, see www.iste.co.uk/lobry/multivariate.zip*

NOTES ON FIGURE 4.3.– The eigenvalue scree plots for nine analyses associated with ICA for structured inertia. This is shown in this figure by subtracting the inertia expected under the null hypothesis H_0 indicated by the red points in Figure 4.2. All figures use the same y-scale to allow for comparisons. Conceptually assigned factors are in color, greedy or unassigned factors in black and unkept factors are in gray. Colors are defined in Figure I.1. ℝ code available in Appendix 1, section A1.5.3.

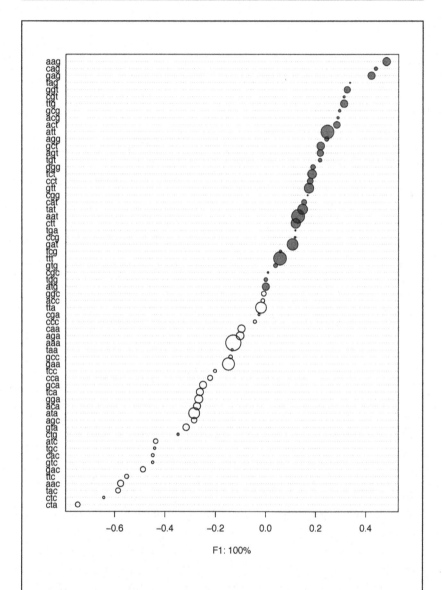

Figure 4.4. *Between-group synonymous factor*

NOTES ON FIGURE 4.4.– Distribution of codons on the first (and single) factor of synonymous between-group analysis with GT-ending codons in gray and AC-ending codons in white. The area of the circles is proportional to the total number of codons in the data set. ℛ code available in Appendix 1, section A1.5.4.

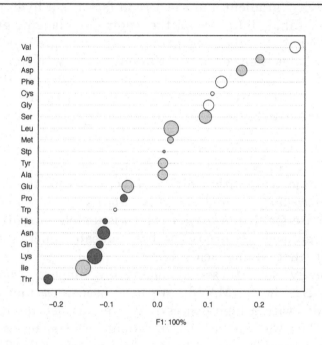

Figure 4.5. *Between-group amino acid factor*

NOTES ON FIGURE 4.5.– Distribution of amino acids on the first (and single) factor of non-synonymous between-group analysis. The area of the circles is proportional to the total number of amino acids in the data set. Amino acids in white are encoded exclusively with codons that have GT bases in first and second positions and those in dark gray have AC bases. ℛ code available in Appendix 1, section A1.5.5.

4.3.2. *Within-group analysis*

4.3.2.1. *Regular factors*

Figure 4.6 shows that the first factor is a "gene expressivity" gradient and the second factor is a hydropathy gradient. Taking into account the leading and lagging groups in the analysis, the degeneracy observed previously in Figure 3.5 has been removed. In a sense we have restored the second factorial map for global CA in Figure 2.8 but the big difference is that we know here that the factors are present at the amino acid level.

For the "gene expressivity" factor, it would be tempting as in [AKA 02] to see if proteins with a high expressivity tend to avoid amino acids with a high metabolic cost for the cell, as do *Escherichia coli* and *Bacillus subtilis*. This would not make sense for *B. burgdorferi*, which is unable to synthesize any amino acids *de novo*. The best educated guess we can make is that amino acids favored in genes with a high expressivity are those abundant by import from *B. burgdorferis*' hosts [MA 18]. We have seen in section 2.5.2 that genes with a high expressivity are enriched in GC-rich codons, but we are now in the amino acid space, so the question is how to translate this. The model [LOB 97] that predicts the frequency of a given amino acid aa as a function of the GC content θ in the absence of selective constraints is defined by:

$$P(\theta, aa) = \frac{f(\theta, aa)}{8 - (1 - \theta)^2(1 + \theta)} \qquad [4.1]$$

with $\theta \in [0, 1]$ and

$$
f(\theta, aa) = \begin{cases}
(1 - \theta)^2(2 - \theta) & \text{if } aa \in \{\text{Ile}\} \\
(1 - \theta)^2 & \text{if } aa \in \{\text{Phe, Lys, Tyr, Asn}\} \\
1 - \theta^2 & \text{if } aa \in \{\text{Leu}\} \\
(1 - \theta)^2\theta & \text{if } aa \in \{\text{Met}\} \\
(1 - \theta)\theta & \text{if } aa \in \{\text{Asp, Glu, His, Gln, Cys}\} \\
2(1 - \theta)\theta & \text{if } aa \in \{\text{Val, Thr}\} \\
3(1 - \theta)\theta & \text{if } aa \in \{\text{Ser}\} \\
(1 - \theta)\theta^2 & \text{if } aa \in \{\text{Trp}\} \\
\theta(\theta + 1) & \text{if } aa \in \{\text{Arg}\} \\
2\theta^2 & \text{if } aa \in \{\text{Gly, Pro, Ala}\} \quad [4.2]
\end{cases}
$$

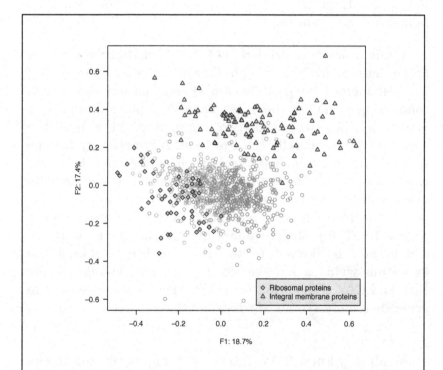

Figure 4.6. *Within-group amino acid factors. For a color version of this figure, see www.iste.co.uk/lobry/multivariate.zip*

NOTES ON FIGURE 4.6.– First factorial map, (F_1, F_2), for within-group non-synonymous codon usage analysis. The colors defined in Figure I.1 are used for sequences coding for integral membrane proteins and ribosomal proteins. ® code available in Appendix 1, section A1.5.6.

This is a simple probabilistic model in which coding sequences are generated by random sampling from a DNA urn with a given GC content. The numerator reflects the structure of the genetic code and the denominator is a correcting factor due to stop codons. Eight classes of response to GC (see [LOB 97] for details) is used to define the gray scale with amino acids favored by high GC darker in Figure 4.7. Gene with a high expressivity are thus enriched in high GC amino acids.

As the promise in section 2.6.1 is a debt that we may not forget, let see now how we can illustrate in a simple way that integral membrane proteins are shorter on average. For a continuous variable, the most popular representation is a histogram, but we need two histograms here and the superposition is usually hard to read. A simple way to cope with this is to use modern tools such as the kernel density estimates provided by the standard ® density() function. To avoid scaring the reader unfamiliar with these tools, it is possible to plot a regular histogram in the background as in Figure 4.8. The problem we face here is that the protein size distribution is skewed to the right and that taking the logarithm yields a left-skewed distribution. Thanks to the BOX and COX transformation [BOX 64] it is possible here to generate a more symmetric distribution.

4.3.2.2. Greedy factors

We already know from section 2.6.2 that we are too greedy here in keeping those factors for interpretation. This section is therefore rated X and a parental consent may be required in some countries before further reading. Figure 4.9 shows that the factors are essentially defined by the two rarest (see Figure 2.3) amino acids Trp and Cys. Note that the pseudo-amino acid Stp also appears here, and as discussed in section 2.1.2, this is an indicator of a protein size effect. Adding protein size as an indicative variable, we note that most outliers are small proteins. Our greediness is hardly punished by the triviality of the interpretation: it is easier to have by pure chance a relatively high concentration in a rare amino acid when you are a small protein.

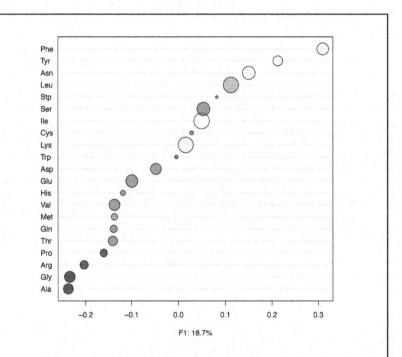

Figure 4.7. *Within-group amino acid first factor*

NOTES ON FIGURE 4.7.– Coordinates of amino acids on the first factor for the within-group amino acid usage. Amino acids encoded by GC-rich codons are darker. ®️ code available in Appendix 1, section A1.5.7.

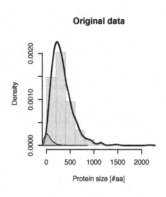

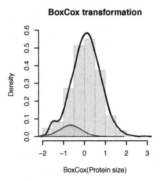

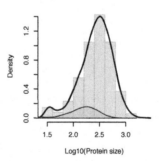

Figure 4.8. *Integral membrane proteins are smaller. For a color version of this figure, see www.iste.co.uk/lobry/multivariate.zip*

NOTES ON FIGURE 4.8.– The colors defined in Figure I.1 are used for integral membrane proteins. The top panel shows that the original distribution of protein size is skewed to the right so that it is difficult to read the graph. The logarithmic transformation in the bottom panel improves the situation but the distribution is now skewed to the left. The BOX and COX transformation [BOX 64] in the middle panel (here with $\lambda = 0.285$) is a transformation intermediate between the top and bottom panels, to normalize data to the best, and then improve the reading of the graph. ℝ code available in Appendix 1, section A1.5.8.

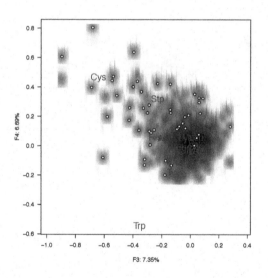

Figure 4.9. *An X-rated factorial map. For a color version of this figure, see www.iste.co.uk/lobry/multivariate.zip*

NOTES ON FIGURE 4.9.– Protein and amino acid location on the $F_3 - F_4$ factorial map for within-group amino acid usage. White circles are proteins with less than 75 amino acids. ℝ code available in Appendix 1, section A1.5.9.

Conclusion

There is something fascinating about starting from a simple table of integer values and a few illustrative variables we end with so many biological factors. Here is a summary with references to relevant sections.

1) A symmetric directional mutation pressure that depletes all coding sequences from GC-rich codons (section 2.3), which is strong enough to modulate the amino acid content of the encoded protein (section 3.3.1).

2) An asymmetric directional mutation pressure (section 2.4), which is a kind of dictatorial factor at the synonymous level (section 3.2.1) and strong enough to modulate the amino acid content of the encoded protein (section 4.3.1).

3) A selective pressure to avoid head-on collisions between the RNA polymerase and the replication fork (section 2.5.3).

4) A selective pressure related to "gene expressivity" (section 2.5), which is visible only at the amino acid level (section 4.3.2.1).

5) A selective pressure to maintain the subcellular the location of integral membrane proteins (section 2.5) by increasing their content in hydrophobic amino acids (section 4.3.2.1).

6) A selective pressure that prevents the size of the integral membrane protein being too large, so that it can fit into the membrane (sections 2.6.1 and 4.3.2.1).

Appendices

Appendix 1

A1.1. Introduction

A1.1.1. *Code for Figure I.1*

```
collealag <- c(rgb(0.1, 0.1, 0.8), rgb(0.8, 0.1, 0.1)) # leading in
        blue, lagging in red
collea <- col2alpha(rgb(0.1, 0.1, 0.9)) ; pchlea <- 21
collag <- col2alpha(rgb(0.9, 0.1, 0.1)) ; pchlag <- 22
colrib <- col2alpha(rgb(0.2, 0.9, 0.2)) ; pchrib <- 23
colimp <- col2alpha("cyan") ; pchimp <- 24
par(mfrow = c(2, 2), mar = c(0, 0, 3, 0))
myplot <- function(cols, main, ...){
  plot(0, type = "n", ann = FALSE, xaxt = "n", yaxt = "n",
          bty = "n", ...)
  legend("top", legend = c("leading", "lagging", "ribosomal",
    "integral membrane protein"), pch = c(pchlea, pchlag, pchrib,
          pchimp),
    pt.bg = cols, cex = 1.75, box.lwd = 0, box.col = "transparent",
          xpd = NA)
  title(main = main, cex.main = 1.5)
}
cols <- c(collea, collag, colrib, colimp)
myplot(cols, "Normal")
library(dichromat)
myplot(dichromat(cols, "deutan"), "Deutan")
myplot(dichromat(cols, "protan"), "Protan")
myplot(dichromat(cols, "tritan"), "Tritan")
```

A1.1.2. *Code for Figure I.2*

```
library(MASS)
data(survey)
survey.cc <- survey[complete.cases(survey), ]
col <- ifelse(survey.cc$Sex == "Male", collea, collag)
pch <- ifelse(survey.cc$Sex == "Male", pchlea, pchlag)
df <- survey.cc[ , c("Wr.Hnd", "NW.Hnd", "Height")]
plot(df, col = col2alpha("grey"), bg = col, pch = pch, las = 1)
library(rgl)
dfs <- scale(df)
plot3d(dfs, type = "s", col = col, box = FALSE, xlab = "Wr.Hnd",
        ylab = "NW.Hnd",
  zlab = "Height", zoom = 5)
plot3d(ellipse3d(cor(dfs)), type = "wire", add = TRUE)
plot3d(dfs, type = "s", col = col, box = FALSE, xlab = "", ylab = "",
  zlab = "", axes = FALSE)
plot3d(ellipse3d(cor(dfs)), type = "wire", add = TRUE)
umat <- matrix(c(
0.72723716, 0.5168677, 0.4516346, 0,
-0.09930068, -0.5718403, 0.8143328, 0,
0.67916495, -0.6370609, -0.3645386, 0,
0.00000000, 0.0000000, 0.0000000, 1), nrow = 4, ncol = 4)
rgl.viewpoint(userMatrix = t(umat), type = "modelviewpoint")
mklab <- function(x, n){
  paste("F", n, ": ", signif(100*x$eig[n]/sum(x$eig), 3), "%", sep = "")
}
pca <- dudi.pca(df, scannf = FALSE)
x <- pca$li[ , 1] ; y <- pca$li[ , 2]
xlab <- mklab(pca, 1) ; ylab <- mklab(pca, 2)
plot(x, y, col = col2alpha("grey"), bg = col, pch = pch, asp = 1,
        xlab = xlab, ylab = ylab)
```

A1.1.3. *Code for Figure I.3*

```
numcode <- 1            # To choose the standard genetic code
pi <- 3.14159265358979 # Better than God's approximation
par(mar = c(0, 0, 0, 0))
symbols(x = rep(0,3), y = rep(0,3), circles = c(1, 0.75, 0.45),
   inches = FALSE, bg = c("pink", "white", "lightblue"), xlim = c(-1, 1),
   ylim = c(-1, 1), bty = "n", asp = 1, xlab = "", ylab= "", xaxt ="n",
          yaxt = "n")
words() -> codons
unlist(lapply(lapply(codons,s2c),translate, numcode = numcode)) -> aa
aaa(aa) -> aa3
neworder <- order(aa3)
aa3 <- aa3[neworder]
aa <- aa[neworder]
codons <- codons[neworder]
cangles <- seq(0, 2*pi, length.out = 65)[1:64]
text(x = sin(cangles)*0.9, y = cos(cangles)*0.9, labels =
          toupper(codons), cex = 0.7)
aangles <- seq(0, 2*pi, le = 22)[1:21]
text(x = sin(aangles)*0.35, y = cos(aangles)*0.35, labels = unique(aa3),
          cex = 0.8)
text(x = sin(aangles)*0.25, y = cos(aangles)*0.25, labels = unique(aa),
          cex = 0.8)
for( i in 1:64 )
{
   target <- aaa(translate(s2c(codons[i]), numcode = numcode))
   n <- which( unique(aa3) == target)
   lines(x = c(sin(cangles[i])*0.85, sin(aangles[n])*0.4),
         y = c(cos(cangles[i])*0.85, cos(aangles[n])*0.4) )
}
```

A1.1.4. *Code for Figure I.4*

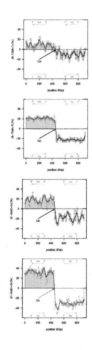

```
if(final.query0){
  choosebank("emglib")
  borre <- query("borre","n=BORBUCG")
  bbg <- getSequence(borre$req[[1]], as.string = TRUE)
  cds <- query("cds", "borre and t=CDS")
  n <- cds$nelem
  pos <- matrix(NA, nrow = n, ncol = 2) # CDS positions on chromosome
  for(i in 1:n){
    tmp <- getAnnot(cds$req[[i]], nbl = 1)
    tmp2 <- c2s(s2c(tmp)[s2c(tmp) %in% c(0:10, ".")])
    tmp2 <- unlist(strsplit(tmp2, split = "\\."))
    if(length(grep("complement", tmp)) == 1){
      pos[i, 1] <- as.numeric(tmp2[3])
      pos[i, 2] <- as.numeric(tmp2[1])
    } else {
      pos[i, 1] <- as.numeric(tmp2[1])
      pos[i, 2] <- as.numeric(tmp2[3])
    }
  }
  save(bbg, pos, file = "local/bbg.rda")
  closebank()
} else {
  load("local/bbg.rda")
}
skews <- function(x){
  if( !is.character(x) || length(x) > 1 ) stop("single string expected")
```

```r
    tmp <- tolower(s2c(x))
    nC <- sum(tmp == "c") ; nG <- sum(tmp == "g")
    if(nC + nG == 0) gcskew <- NA else gcskew <- 100*(nC - nG)/(nC + nG)
    nA <- sum(tmp == "a") ; nT <- sum(tmp == "t")
    if(nA + nT == 0) atskew <- NA else atskew <- 100*(nA - nT)/(nA + nT)
    return(list(gcskew = gcskew, atskew = atskew))
}
step <- 2000 ; wsize <- 10000
starts <- seq(from = 1, to = nchar(bbg), by = step)
starts <- starts[-length(starts)] # remove last one
n <- length(starts)
# All positions
atskew <- gcskew <- numeric(n)
for(i in seq_len(n)){
  tmp <- skews(substr(bbg, starts[i], starts[i] + wsize - 1))
  gcskew[i] <- tmp$gcskew
  atskew[i] <- tmp$atskew
}
myplot <- function(x, y, ...){
  plot(x,y, col = "grey", type = "b", ylim = c(-55, 55), las = 1,
          xaxt = "n",
  xlab = "position (Kbp)", ...)
  axis(1, at = seq(0, 1000, by = 200))
  lines(smooth <- lowess(x, y, f = 0.05))
  polycurve <- function(x, y, base.y = min(y), ...)
    polygon(x = c(min(x), x, max(x)), y = c(base.y, y, base.y), ...)
  up <- smooth$y > 0
  polycurve(smooth$x[up], smooth$y[up], base.y = 0,
          col = col2alpha("grey"))
  abline(h = 0)
  arrows(200, -20, 450, -1, length = 0.1, angle = 20)
  text(200, -20, "Ori", pos = 1)
  myarrows <- function(x, lea){
    col <- ifelse(lea, collea, collag)
    if(x < 500) lea <- !lea
    y <- ifelse(lea, -50, 50)
    pos <- ifelse(lea, 1, 3)
    code <- ifelse(lea, 2, 1)
    arrows(x, y, x + 200, y, col = col, code = code, length = 0.1,
            angle = 20)
    text(x + 100, y, ifelse(col == collea, "lea", "lag"), col = col,
            pos = pos)
    text(ifelse(lea, x, x + 200), y, "5'", col = col, pos = pos)
    text(ifelse(lea, x + 200, x), y, "3'", col = col, pos = pos)
  }
  for(a in c(100, 600)) for(b in c(TRUE, FALSE)) myarrows(a, b)
}
par(mfrow = c(2, 2), mar = c(5, 4, 1, 0) + 0.1)
myplot(starts/1000, gcskew, ylab = "(C-G)/(C+G) [%]")
myplot(starts/1000, atskew, ylab = "(A-T)/(A+T) [%]")
# Third codon positions
bbg3 <- rep("n", nchar(bbg))
bbgc <- s2c(bbg)
for(i in 1:nrow(pos)){
  if(pos[i, 1] < pos[i, 2]){ # direct strand
    tcp <- seq(pos[i, 1] + 2, pos[i, 2], by = 3)
  } else { # complementary strand
    tcp <- seq(pos[i, 1] - 2, pos[i, 2], by = -3)
  }
  bbg3[tcp] <- bbgc[tcp]
```

```
}
bbg3 <- c2s(bbg3)
atskew <- gcskew <- numeric(n)
for(i in seq_len(n)){
  tmp <- skews(substr(bbg3, starts[i], starts[i] + wsize - 1))
  gcskew[i] <- tmp$gcskew
  atskew[i] <- tmp$atskew
}
myplot(starts/1000, gcskew, ylab = "(C-G)/(C+G) [%]")
myplot(starts/1000, atskew, ylab = "(A-T)/(A+T) [%]")
```

A1.2. Chapter 1

A1.2.1. *Code for Figure 1.1*

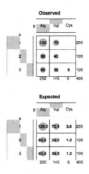

```
data(toyaa)
ct <- t(as.table(as.matrix(toyaa)))
library(gplots)
par(mfrow = c(1,2), mar = c(0, 1, 1, 1) + 0.1)
balloonplot(ct, main = "Observed", xlab = "y", ylab = "x")
balloonplot(as.table(chisq.test(ct)$expected), main = "Expected",
        xlab = "y", ylab = "x")
```

A1.2.2. *Code for Figure 1.2*

```
nsimulations <- 10000
set.seed(1)
r <- rowSums(toyaa) ; c <- colSums(toyaa)
theo <- (r %*% t(c))/sum(toyaa)
```

```
f <- function(){
  obs <- r2dtable(1, r, c)[[1]]
  return( sum((obs - theo)^2/theo))
}
chi2.obs <- sum((toyaa - theo)^2/theo)
sim <- replicate(nsimulations, f())
par(mar = c(4, 4, 1, 0) + 0.1)
hist(sim, probability = TRUE, col = "lightblue", main = "", las = 1,
  xlab = expression(chi^2*plain("-statistic")))
arrows(x0 = chi2.obs, y0 = 0.05, x1 = chi2.obs, y1 = 0, angle = 10,
       length = 0.15)
text(chi2.obs, y = 0.05, labels = expression(chi[obs]^2), pos = 3)
xx <- seq(0, max(sim, chi2.obs), length.out = 255)
lines(xx, dchisq(xx, df = 4), xpd = NA)
```

A1.2.3. *Code for Figure 1.3*

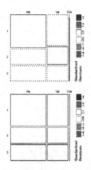

```
par(mfrow = c(1, 2), mar = c(0, 0, 0, 0) + 0.5)
mosaicplot(t(toyaa), shade = TRUE, main = "", las = 1)
mosaicplot(chisq.test(t(toyaa))$expected, shade = TRUE, main = "",
       las = 1)
```

A1.2.4. *Code for Figure 1.4*

```
cex <- 2 ; pch <- 21:23 ; bg <- c("black", "cyan", "magenta")
myplot <- function(res, panel, ... )
{
```

```
    plot(res$li[ , 1], res$li[ , 2], asp = 1, main = "", xlab = "",
        ylab = "", cex = cex, pch = pch, bg = bg, ...)
    text(x = res$li[ , 1], y = res$li[ , 2], labels = 1:3,
        pos = ifelse(res$li[ , 2] < 0, 1, 3))
    perm <- c(3, 1, 2)
    lines( c(res$li[ , 1], res$li[perm, 1]), c(res$li[ , 2],
        res$li[perm, 2]))
    points(res$li[ , 1], res$li[ , 2], pch = pch, bg = bg, cex = cex)
    legend("topleft", legend = panel, bty = "n", text.font = 2, cex = 1.5)
}
par(mfrow = c(2, 2), mar = c(2, 2, 1, 0) + 0.1)
# Panel (1, 1)
pco <- dudi.pco(dist(toyaa), scann = FALSE, nf = 2)
myplot(pco, "a")
# Panel (1, 2)
profile <- toyaa/rowSums(toyaa)
dudi.pco(dist(profile), scann = F, nf = 2) -> pco1
myplot(pco1, "b")
# Panel (2, 1)
doubleprofile <- t(t(profile)/(colSums(toyaa)))
dudi.pco(dist(doubleprofile), scann = F, nf = 2) -> pco1
myplot(pco1, "c")
# Panel (2, 2)
ca <- sum(toyaa)*t(t(profile)/(sqrt(colSums(toyaa))))
dudi.pco(dist(ca), scann = F, nf = 2) -> pco1
myplot(pco1, "d")
```

A1.2.5. *Code for Figure 1.5*

```
library(MASS)  # for mvrnorm()
set.seed(1)    # for reproducibility
n <- 500       # number of points in each groupe
Sigma <- matrix(c(1, 0.99, 0.99, 1), nrow = 2, ncol = 2)
df <- mvrnorm(n = n, mu = c(0,  0), Sigma)
par(mfcol = c(3, 3), mar = rep(0.1, 4), oma = rep(1, 4))
myplot <- function(df, shift){
  x <- rbind(df, df + rep(c(shift, -shift), each = n))
  plot(x, pch = 19, cex = 0.5, ann = FALSE, xaxt = "n", yaxt = "n")
  res <- dudi.pca(x, scan = FALSE, scale = FALSE)
  scatter(res, clab.row = 0, clab.col = 0, col = "grey",
        posieig = "none")
  screeplot(res, npcs = 10, main = "", ylim = c(0, 6), yaxt = "n")
  box()
}
mytext <- function(...) text(..., pos = 4, cex = 2)
txt1 <- "Size effect"
txt2 <- "Between groups\neffect"
```

```
par(lheight = 0.75)
myplot(df, 1)
mytext(0, 2.5, txt1) ; mytext(1, 1, txt2)
myplot(df, 2.011269)
mytext(0, 2.5, "?") ; mytext(1, 2.5, "?")
myplot(df, 3)
mytext(1, 2.5, txt1) ; mytext(0, 5, txt2)
```

A1.2.6. *Code for Figure 1.6*

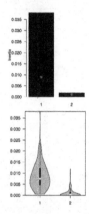

```
myscreeplot <- function(x, y, npcs = 15, nf = x$nf,
                        p.pch = 19, p.col = "red", p.cex = 0.75, ...){
    screeplot(x, npcs = npcs, ...) ; res.barplot <- barplot(x$eig,
            plot = FALSE)
    imax <- min(npcs, x$rank)
    points(res.barplot[1:imax, 1], y[1:imax], pch = p.pch, col = p.col,
            cex = p.cex)
}
par(mfrow = c(1, 2), mar = c(2, 4, 1, 1) + 0.1)
nreplicates <- 250
set.seed(1) # for reproducibility
toyaa.ca <- dudi.coa(toyaa, scannf = FALSE)
rs <- rowSums(toyaa) ; cs <- colSums(toyaa)
f <- function(){
    rtable <- r2dtable(1, rs, cs)[[1]]
    eig <- dudi.coa(rtable, scannf = FALSE)$eig
    length(eig) <- min(dim(toyaa)) - 1
    return(eig)
}
sim <- replicate(nreplicates, f())
sim[which(is.na(sim), arr.ind = TRUE)] <- 0
ymax <- toyaa.ca$eig[1]
myscreeplot(toyaa.ca, rowMeans(sim), 2, las = 1, main = "",
        ylim = c(0, ymax), p.cex = 1.25)
library(vioplot)
resbxp <- barplot(toyaa.ca$eig, las = 1, main = "", ylim = c(0, ymax),
    col = "transparent", border = "transparent")
```

```
for(i in 1:nrow(sim)) vioplot(sim[i, ], at = resbxp[i, 1], add = TRUE,
        col = "lightblue")
axis(1, at = resbxp[, 1], labels = 1:nrow(sim))
```

A1.2.7. *Code for Figure 1.7*

```
data(toyaa)
profile <- toyaa/rowSums(toyaa)
ca <- sum(toyaa)*t(t(profile)/(sqrt(colSums(toyaa))))
myplot <- function(res, ... )
{
  plot(res$li[ , 1], res$li[ , 2], ...)
  text(x = res$li[ , 1], y = res$li[ , 2], labels = 1:3,
        pos = ifelse(res$li[ , 2] < 0, 1, 3))
  perm <- c(3, 1, 2)
  lines( c(res$li[ , 1], res$li[perm, 1]), c(res$li[ , 2],
        res$li[perm, 2]))
}
par(mfrow = c(2, 2), mar = c(0, 0, 0, 0) + 0.0, xaxt = "n", yaxt = "n",
    oma = c(1, 1, 1, 1))
dudi.pco(dist(ca), scann = F, nf = 2) -> pco
cex <- 2 ; pch <- 21:23 ; bg <- c("black", "cyan", "magenta")
myplot(pco, main = "", asp = 1, pch = pch, xlab = "", ylab = "",
        cex = cex, bg = bg)
pco$li[, 1] <- -1*pco$li[, 1]
myplot(pco, main = "", asp = 1, pch = pch, xlab = "", ylab = "",
        cex = cex, bg = bg)
pco$li[, 2] <- -1*pco$li[, 2] ; pco$li[, 1] <- -1*pco$li[, 1]
myplot(pco, main = "", asp = 1, pch = pch, xlab = "", ylab = "",
        cex = cex, bg = bg)
pco$li[, 1] <- -1*pco$li[, 1]
myplot(pco, main = "", asp = 1, pch = pch, xlab = "", ylab = "",
        cex = cex, bg = bg)
```

A1.2.8. *Code for Figure 1.8*

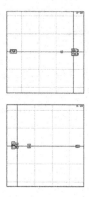

```
toyaa.ca <- dudi.coa(toyaa, scannf = FALSE)
ttoyaa.ca <- dudi.coa(t(toyaa), scannf = FALSE)
par(oma = c(0, 0, 0, 0) + 1)
layout(matrix(c(1, 0, 2), nrow = 1, ncol = 3, byrow = TRUE),
       widths = c(3, 0.5, 3))
scatter(toyaa.ca, posieig = "none")
scatter(ttoyaa.ca, posieig = "none")
```

A1.2.9. *Code for Figure 1.9*

```
delta <- 1
x <- seq(-4, 4, length.out = 255)
par(mfrow = c(2, 2), oma = c(0, 0, 0, 0) + 0.2, mar = c(2, 2, 2, 2))
# Panel a
plot(x, dnorm(x), type = "l", xlab = "gradient", las = 1,
     ylab = "Aminoacid concentration")
lines(x, dnorm(x, mean = -delta), lty = 2)
lines(x, dnorm(x, mean = delta), lty = 4)
legend("topright", inset = 0.02, legend = c("Ala", "Val", "Cys"),
       lty = c(2, 1, 4))
legend("topleft", legend = "a", bty = "n", text.font = 2, cex = 1.5)
x <- seq(-3, 3, length.out = 40)
rug(x, col = "red")
# Panel b
grd <- floor(1000*cbind(dnorm(x, mean = -delta), dnorm(x),
       dnorm(x, mean = delta)))
```

```
grd <- as.data.frame(grd)
colnames(grd) <- c("Ala", "Val", "Cys")
library(plot3D)
points3D(grd[, 1], grd[, 2], grd[, 3], colvar = NULL, pch = 19,
        xlab = "Ala", ylab = "Val", zlab = "Cys", phi = 20, theta = 40)
legend("topleft", legend = "b", bty = "n", text.font = 2, cex = 1.5)
# Panel c
triangle.plot(grd, labeltriangle = TRUE, show.position = FALSE)
legend("topleft", legend = "c", bty = "n", text.font = 2, cex = 1.5)
# Panel d
scatter(dudi.coa(grd, scannf = FALSE), posieig = "none", grid = FALSE)
legend("topleft", legend = "d", bty = "n", text.font = 2, xpd = NA,
        cex = 1.5)
```

A1.2.10. *Code for Figure 1.10*

```
mklab <- function(x, n){
  paste("F", n, ": ", signif(100*x$eig[n]/sum(x$eig), 3), "%", sep = "")
}
if(final.stability){
  load("local/bact.rda")
  tdf <- t(bact[, 4:67]) # codon count colums
  tdf.coa <- dudi.coa(tdf, scannf = FALSE, nf = 2)
  facaa <- as.factor(aaa(sapply(rownames(tdf),
          function(x) translate(s2c(x)))))
  tdf.wca <- wca(tdf.coa, facaa, scan = FALSE, nf = 2)
  # Symmetry choice
  tdf.wca$li[, 1] <- -1*tdf.wca$li[, 1] ;
          tdf.wca$co[, 1] <- -1*tdf.wca$co[, 1]
  tdf.wca$li[, 2] <- -1*tdf.wca$li[, 2] ;
          tdf.wca$co[, 2] <- -1*tdf.wca$co[, 2]
  tdf.bca <- bca(tdf.coa, facaa, scan = FALSE, nf = 2)
  # Symmetry choice
  tdf.bca$li[, 2] <- -1*tdf.bca$li[, 2] ; tdf.bca$co[, 2]
          <- -1*tdf.bca$co[, 2]
  topt <- read.table(
    "http://pbil.univ-lyon1.fr/members/lobry/repro/gene06/toptsummary.
            january.table",
    header = TRUE, stringsAsFactors = FALSE)
  save(topt, bact, tdf, tdf.coa, tdf.wca, tdf.bca,
          file = "local/stabilty.rda")
  } else {
  load("local/stabilty.rda")
  }
f <- function(x){
  if(x[1] == "CANDIDATUS")
    return(paste(x[2:3], collapse = " "))
```

```
    else
      return(paste(x[1:2], collapse = " "))
}
gensp <- sapply(strsplit(bact$species, split = " "), f)
genus <- sapply(strsplit(gensp, split = " "), function(x) x[1])
showsp <- function(x, who, ...){
  sel <- x[gensp == who, 1:2]
  id <- chull(sel)
  polygon(sel[id, 1], sel[id, 2], ...)
}
par(mar = c(4, 4, 1, 0) + 0.1, mfrow = c(2, 2))
# Synonymous codon usage in sequence space
xlab <- mklab(tdf.wca, 1) ; ylab <- mklab(tdf.wca, 2)
smoothScatter(tdf.wca$co[, 1], tdf.wca$co[, 2], xlab = xlab,
          ylab = ylab,
    colramp = colorRampPalette(c("white", "black")), nrpoints = 0,
    xlim = c(-1, 1), ylim = c(-0.4, 0.6))
showlist <- c("STAPHYLOCOCCUS AUREUS", "ENTEROCOCCUS FAECALIS",
"ESCHERICHIA COLI", "SALMONELLA ENTERICA", "KLEBSIELLA PNEUMONIAE",
"MYCOBACTERIUM TUBERCULOSIS")
for(s in showlist) showsp(tdf.wca$co,s, col = "white")
hyp <- gensp %in% topt[topt$toptclass == "hyperthermophile", "species"]
for(i in which(hyp)) points(tdf.wca$co[i, 1], tdf.wca$co[i, 2], pch = 21,
          bg = col2alpha("red"),
    col = col2alpha(grey(0.5)))
# Synonymous codon usage in codon space
cex <- 50*sqrt(tdf.wca$lw)
bg <- ifelse(substring(rownames(tdf.wca$tab), 3, 4) %in% c("a", "t"),
    "transparent", col2alpha(grey(0.2)))
plot(tdf.wca$li[,1], tdf.wca$li[,2], xlab = xlab, ylab = ylab,
    xlim = c(-1.2, 1), ylim = c(-0.4, 0.6), cex = cex, xpd = NA, pch = 21,
    bg = bg, col = col2alpha(grey(0.5)))
text(tdf.wca$li[,1], tdf.wca$li[,2], rownames(tdf.wca$tab), cex = 0.5)
# Aminoacid usage in protein space
xlab <- mklab(tdf.bca, 1) ; ylab <- mklab(tdf.bca, 2)
smoothScatter(tdf.bca$co[,1], tdf.bca$co[,2], xlab = xlab, ylab = ylab,
    xlim = c(-0.5, 0.5), ylim = c(-0.2, 0.2), nrpoints = 0,
    colramp = colorRampPalette(c("white", "black")))
for(s in rev(showlist)) showsp(tdf.bca$co,s, col = "white")
for(i in which(hyp)) points(tdf.bca$co[i, 1], tdf.bca$co[i, 2], pch = 21,
          bg = col2alpha("red"),
    col = col2alpha(grey(0.5)))
# Aminoacid usage in aa space
cex <- 25*sqrt(tdf.bca$lw)
bgaagc <- function(aa){
  bg <- rep(col2alpha(grey(0.5)), length(aa))
  bg[aa == "Ile"] <- col2alpha(grey(1))
  bg[aa %in% c("Phe", "Lys", "Tyr", "Asn")] <- col2alpha(grey(0.95))
  bg[aa == "Leu"] <- col2alpha(grey(0.7))
  bg[aa == "Met"] <- col2alpha(grey(0.6))
  bg[aa == "Trp"] <- col2alpha(grey(0.4))
  bg[aa == "Arg"] <- col2alpha(grey(0.05))
  bg[aa %in% c("Gly", "Pro", "Ala")] <- col2alpha(grey(0))
  return(bg)
}
plot(tdf.bca$li[,1], tdf.bca$li[,2], xlab = xlab, ylab = ylab,
    xlim = c(-0.5, 0.5), ylim = c(-0.2, 0.2), cex = cex, xpd = NA,
          pch = 21,
    bg = bgaagc(rownames(tdf.bca$li)), col = col2alpha(grey(0.5)))
text(tdf.bca$li[,1], tdf.bca$li[,2], rownames(tdf.bca$tab), cex = 0.5)
```

A1.2.11. *Code for Figure 1.11*

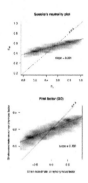

```
load("local/stabilty.rda")
oddcodons <- c("taa", "tag", "tga", "att", "atc", "ata", "atg", "tgg")
dfs <- t(as.matrix(tdf[!rownames(tdf) %in% oddcodons, ]))
p3 <-  sapply(colnames(dfs), function(x) GC3(s2c(x)))
p12 <- sapply(colnames(dfs), function(x) (GC1(s2c(x)) + GC2(s2c(x)))/2)
P3 <- (dfs/rowSums(dfs)) %*% p3
P12 <- (dfs/rowSums(dfs)) %*% p12
x <- P3[ , 1] ; y <- P12[ , 1]
par(mfrow = c(1, 2))
# Left panel
colramp <- colorRampPalette(c("white", "black"))
smoothScatter(x, y, xlim = c(0,1), ylim = c(0,1), colramp = colramp,
    nrpoints = 0, xlab = expression(P[1]), ylab = expression(P[12]),
    las = 1, main = "Sueoka's neutrality plot")
abline(c(0, 1), lty = 2)
w <- as.data.frame(cbind(x, y)) ; cov <- var(w)
u <- eigen(cov, symmetric = TRUE)$vectors[ , 1]
p <- u[2]/u[1]
scal <- (x - mean(x))*u[1] + (y - mean(y))*u[2]
abline( c(mean(y) -p*mean(x), p))
text(0.7, 0.3, paste("slope =", signif(p, 3)))
text(0.9, 0.9, "y = x", srt = 45, pos = 2)
# Rigth panel
x <- tdf.wca$co[, 1] ; y <- tdf.bca$co[, 1]
xlab <- "Strain coordinate on synonymous factor"
ylab <- "Strain coordinate on nonsynonymous factor"
smoothScatter(x, y, asp = 1, colramp = colramp, nrpoints = 0,
    las = 1, main = "First factor (GC)", xlab = xlab, ylab = ylab)
abline(c(0, 1), lty = 2)
w <- as.data.frame(cbind(x, y)) ; cov <- var(w)
u <- eigen(cov, symmetric = TRUE)$vectors[ , 1]
p <- u[2]/u[1]
scal <- (x - mean(x))*u[1] + (y - mean(y))*u[2]
abline( c(mean(y) -p*mean(x), p))
text(0.8, 0.8, "y = x", srt = 45, pos = 2)
text(0.5, -0.5, paste("slope =", signif(p, 3)))
```

A1.3. Chapter 2

A1.3.1. *Code for Figure 2.2*

```
load("local/gc.rda")
if(final.query4){
  choosebank("hogenom7dna")
  tick <- query("tick", "SP=Ixodes scapularis AND T=CDS AND NO
        K=PARTIAL")
  n <- tick$nelem # total number of CDS
  tick.gc <- numeric(n)
  verbose <- TRUE
  for(i in 1:n){
    if(verbose) print(paste("Dealing with sequence number:", i))
    myseq <- getSequence(tick$req[[i]])
    tick.gc[i] <- GC(myseq)
  }
  save(tick.gc, file = "local/tick.rda")
} else {
  load("local/tick.rda")
}
par(mar = c(4, 4, 1, 0) + 0.1)
hist(gc, xlim = c(0, 1), col = grey(0.95), border = grey(0.8), las = 1,
  xlab = "GC-content: (G + C)/(A + T + G + C)", main = "",
        probability = TRUE)
lines(density(gc), lwd = 2)
arrows(0.135, 8, 0.749, 8, length = 0.1, angle = 20, code = 3)
arrows(0.208, 7, 0.748, 7, length = 0.1, angle = 20, code = 3)
lines(density(tick.gc), lty = 2)
par(font = 3)
legend("topright", inset = 0, legend = c("Borrelia burgdorferi",
        "Ixodes scapularis"),
  lwd = c(2, 1), lty = c(1, 2))
```

A1.3.2. *Code for Figure 2.3*

```
par(mar = c(5, 4, 1, 1))
dotchart.uco(colSums(tuco), xlab = "Codon (white) or amino-acid (black)
        count",
  pt.cex = 1, gpch = 19)
abline(v = 0, col = grey(0.9))
```

A1.3.3. *Code for Figure 2.4*

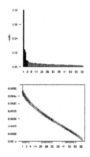

```
if(final.screeplot2){
  nreplicates <- 1000
  set.seed(1) # for reproducibility
  rs <- rowSums(tuco) ; cs <- colSums(tuco)
  f <- function(){
    rtable <- r2dtable(1, rs, cs)[[1]]
    eig <- dudi.coa(rtable, scannf = FALSE)$eig
    length(eig) <- min(dim(tuco)) - 1
    return(eig)
  }
  sim <- replicate(nreplicates, f())
  sim[which(is.na(sim), arr.ind = TRUE)] <- 0
  save(nreplicates, sim, file = "local/screeplot2.rda")
} else {
  load("local/screeplot2.rda")
}
par(mfrow = c(1, 2), mar = c(2, 4, 1, 1) + 0.1)
# Left.panel:
```

```
myscreeplot <- function(x, y, npcs = 15, nf = x$nf,
                        p.pch = 19, p.col = "red", p.cex = 0.75, ...){
  screeplot(x, npcs = npcs, ...) ; res.barplot <- barplot(x$eig,
          plot = FALSE)
  imax <- min(npcs, x$rank)
  points(res.barplot[1:imax, 1], y[1:imax], pch = p.pch, col = p.col,
          cex = p.cex)
}
ylim <- c(0, tuco.coa$eig[1])
myscreeplot(tuco.coa, rowMeans(sim), 63, las = 1, main = "", ylim =ylim,
          p.cex = 0.3)
# Right panel:
library(vioplot)
resbxp <- barplot(tuco.coa$eig, las = 1, main = "", ylim = range(sim),
  col = "transparent", border = "transparent", ylab = "")
for(i in 1:nrow(sim)) vioplot(sim[i, ], at = resbxp[i, 1], add = TRUE,
          col = "lightblue",
    pchMed = ".", colMed = "red", rectCol = "transparent")
axis(1, at = resbxp[, 1], labels = 1:nrow(sim))
```

A1.3.4. *Code for Figure 2.5*

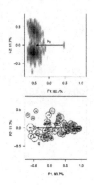

```
mklab <- function(x, n){
  paste("F", n, ": ", signif(100*x$eig[n]/sum(x$eig), 3), "%", sep = "")
}
baby <- rbind(tuco, rep(floor(sum(tuco)/64), 64)) # Adding a uniform
          codon usage CDS
baby.coa <- dudi.coa(baby, scannf = FALSE)
xlab <- mklab(baby.coa, 1) ; ylab <- mklab(baby.coa, 2)
x <- baby.coa$li[, 1] ; xlim <- c(-0.7, 1.1)
y <- baby.coa$li[, 2] ; ylim <- c(-0.8, 0.8)
par(mfrow = c(1, 2), mar = c(4, 4, 1, 1), lend = "butt")
smoothScatter(x, y, las = 1, xlab = xlab, ylab = ylab, xlim = xlim,
          nrpoints = 0,
  ylim = ylim, colramp = colorRampPalette(c("white", "black")))
arrows(0.45, 0, -0.4, 0, lwd = 2, angle = 20, length = 0.1)
text(0, 0, expression(F[0]), pos = 3)
# Right panel
x <- baby.coa$co[, 1]
```

```
y <- baby.coa$co[, 2]
cex <- sqrt(rowSums(tuco))/5
bg <- sapply(colnames(tuco),
    function(x) ifelse(s2c(x)[3] %in% c("c", "g"), col2alpha("grey"),
        "transparent" ))
plot(x, y, las = 1, xlab = xlab, ylab = ylab, cex = cex, xlim = xlim,
        ylim = ylim, pch = 21, bg = bg)
arrows(0.45, 0, -0.4, 0, lwd = 2, angle = 20, length = 0.1, col = "red")
text(x, y, colnames(tuco), cex = 0.5)
arrows(0.45, 0, -0.4, 0, lwd = 2, angle = 20, length = 0.1)
text(0, 0, expression(F[0]), pos = 3)
```

A1.3.5. *Code for Figure 2.6*

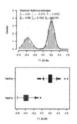

```
mklab <- function(x, n){
    paste("F", n, ": ", signif(100*x$eig[n]/sum(x$eig), 3), "%", sep = "")
}
x <- tuco.coa$li[, 1]
xlab <- mklab(tuco.coa, 1)
xx1 <- seq(min(x), -0.1, length = 200)
xx2 <- seq(-0.1, max(x), length = 200)
mybreaks <- seq(min(x), max(x), length = 20)
logvraineg <- function(param, obs) {
    p <- param[1]
    m1 <- param[2] ; sd1 <- param[3]
    m2 <- param[4] ; sd2 <- param[5]
    -sum(log(p * dnorm(obs, m1, sd1) + (1 - p) * dnorm(obs, m2, sd2)))
}
resnlm <- nlm(f = logvraineg, p = c(0.3, -0.4, 0.1, 0.2, 0.1), obs = x)
estimate <- resnlm$estimate
y1 <- estimate[1]*dnorm(xx1, estimate[2], estimate[3])
y2 <- (1-estimate[1])*dnorm(xx2, estimate[4], estimate[5])
dst <- density(x)
hst <- hist(x, plot = FALSE, breaks = mybreaks, probability = TRUE)
ymax <- 1.2*max(y1, y2, hst$density, dst$y)
par(mfrow = c(1, 2), mar = c(4, 4, 1, 1) + 0.1)
hist(x, col = grey(0.8), probability = TRUE, border = grey(0.9),
    xlab = xlab, main = "", las = 1, breaks = mybreaks, ylim = c(0, ymax))
lines(xx1, y1, col = collealag[2], lwd = 2)
lines(xx2, y2, col = collealag[1], lwd = 2)
text(x = min(x), y = ymax, pos = 4, "Maximum likelihood estimates:")
text(x = min(x), y = 0.9*ymax, col = collealag[2], pos = 4, cex = 1.2,
    substitute(hat(p)[1] == e1~~hat(mu)[1] == e2~~hat(sigma)[1] == e3,
```

```
list(e1 = signif(estimate[1], 3),
    e2 = signif(estimate[2], 3),
    e3 = signif(estimate[3], 3))))
text(x = min(x), y = 0.8*ymax, col = collealag[1], pos = 4, cex = 1.2,
    substitute(hat(p)[2] == q~~hat(mu)[2] == e4~~hat(sigma)[2] == e5,
    list(q = signif(1 - estimate[1], 3),
    e4 = signif(estimate[4], 3),
    e5 = signif(estimate[5], 3))))
boxplot(x~leading, horizontal = TRUE, xlab = xlab, varwidth = TRUE,
    names = c("lagging", "leading"), col = rev(collealag), las = 1,
    pars = list(boxwex = 0.4, staplewex = 0.5, outwex = 0.5))
```

A1.3.6. *Code for Figure 2.7*

```
mklab <- function(x, n){
    paste("F", n, ": ", signif(100*x$eig[n]/sum(x$eig), 3), "%", sep = "")
}
x <- tuco.coa$co[ , 1] # F1 in codon space
y <- colSums(tuco) # Mass of codons
z <- as.data.frame(cbind(x, y))
rownames(z) <- colnames(tuco)
z <- z[order(z$x), ]
pt.cex <- sqrt(z$y)/40
xlab <- mklab(tuco.coa, 1)
color <- ifelse(substr(rownames(z), 3, 3) %in% c("g", "t"),
        col2alpha(grey(0.1)), "transparent")
par(mar = c(5, 3, 1, 1))
dotchart(z$x, labels = rownames(z), pt.cex = pt.cex, bg = color,
        xlab = xlab)
```

A1.3.7. *Code for Figure 2.8*

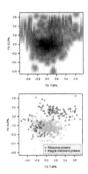

```
mklab <- function(x, n){
  paste("F", n, ": ", signif(100*x$eig[n]/sum(x$eig), 3), "%", sep = "")
}
par(mfrow = c(1, 2), mar = c(5, 4, 0, 0) + 0.5)
x <- tuco.coa$li[, 2] ; y <- tuco.coa$li[, 3]
xlab <- mklab(tuco.coa, 2) ; ylab <- mklab(tuco.coa, 3)
smoothScatter(x, y, xlab = xlab, ylab = ylab, las = 1,
  colramp = colorRampPalette(c("white", "black")), nrpoints = 0)
# Right panel:
plot(x, y, las = 1, xlab = xlab, ylab = ylab, pch = 21, col = grey(0.8))
data(EXP) ; kd <- (tuco/rowSums(tuco)) %*% EXP$KD ; imp <- kd > 0.5
points(x[imp], y[imp], bg = colimp, pch = pchimp)
ribnames <- readLines("local/ribnames.txt")
rib <- rownames(tuco) %in% ribnames
points(x[rib], y[rib], bg = colrib, pch = pchrib)
legend("bottomright", inset = 0.02, c("Ribosomal proteins",
  "Integral membrane proteins"), pch = c(pchrib, pchimp),
  pt.bg = c(colrib, colimp), bg = grey(0.95))
```

A1.3.8. *Code for Figure 2.9*

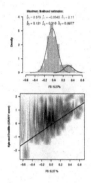

```
mklab <- function(x, n){
  paste("F", n, ": ", signif(100*x$eig[n]/sum(x$eig), 3), "%", sep = "")
}
x <- tuco.coa$li[, 3]
xlab <- mklab(tuco.coa, 3)
xx1 <- seq(-0.5, 0.35, length = 200)
xx2 <- seq(0, 0.6, length = 200)
mybreaks <- seq(min(x), max(x),length = 20)
logvraineg <- function(param, obs) {
  p <- param[1]
  m1 <- param[2] ; sd1 <- param[3]
  m2 <- param[4] ; sd2 <- param[5]
  -sum(log(p * dnorm(obs, m1, sd1) + (1 - p) * dnorm(obs, m2, sd2)))
}
resnlm <- nlm(f = logvraineg, p = c(0.9, 0, 0.1, 0.3, 0.1), obs = x)
estimate <- resnlm$estimate
y1 <- estimate[1]*dnorm(xx1, estimate[2], estimate[3])
y2 <- (1-estimate[1])*dnorm(xx2, estimate[4], estimate[5])
dst <- density(x)
hst <- hist(x, plot = FALSE, breaks = mybreaks)
ymax <- 1.25*max(y1, y2, hst$density, dst$y)
# Left panel:
par(mfrow = c(1, 2), mar = c(4, 4, 1, 1) + 0.1)
hist(x, col = grey(0.8), probability = TRUE, border = grey(0.9),
  xlab = xlab, main = "", las = 1, breaks = mybreaks, ylim = c(0, ymax))
lines(xx1, y1, col = "red", lwd = 2)
lines(xx2, y2, col = "blue", lwd = 2)
text(x = min(x), y = ymax, pos = 4, "Maximum likelihood estimates:")
text(x = min(x), y = 0.9*ymax, col = "red", pos = 4, cex = 1.2,
  substitute(hat(p)[1] == e1~~hat(mu)[1] == e2~~hat(sigma)[1] == e3,
  list(e1 = signif(estimate[1], 3),
    e2 = signif(estimate[2], 3),
    e3 = signif(estimate[3], 3))))
text(x = min(x), y = 0.8*ymax, col = "blue", pos = 4, cex = 1.2,
  substitute(hat(p)[2] == q~~hat(mu)[2] == e4~~hat(sigma)[2] == e5,
  list(q = signif(1 - estimate[1], 3),
    e4 = signif(estimate[4], 3),
    e5 = signif(estimate[5], 3))))
# Right panel figure
xlab <- paste("F3:", signif(100*tuco.coa$eig[3]/sum(tuco.coa$eig), 3),
```

```
        "%")
smoothScatter(x, kd, ylab = "Kyte and Doolittle (GRAVY score)",
        xlab = xlab,
   las = 1, colramp = colorRampPalette(c("white", "black")),
            nrpoints = 0)
abline(lm(kd~x), lwd = 2)
```

A1.3.9. *Code for Figure 2.10*

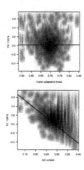

```
y <- tuco.coa$li[, 2]
ylab <- paste("F2:", signif(100*tuco.coa$eig[2]/sum(tuco.coa$eig), 3),
        "%")
par(mfrow = c(1, 2), mar = c(4, 4, 1, 0) + 0.1)
smoothScatter(cai, y, xlab = "Codon adaptation Index", ylab = ylab,
        las = 1,
   main = "", colramp = colorRampPalette(c("white", "black")),
            nrpoints = 0)
abline(lm(y~cai), lwd = 2)
smoothScatter(gc, y, xlab = "GC content", ylab = ylab, las = 1,
        main = "",
   colramp = colorRampPalette(c("white", "black")), nrpoints = 0)
abline(lm(y~gc), lwd = 2)
```

A1.3.10. *Code for Figure 2.11*

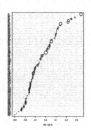

```
x <- tuco.coa$co[ , 2] # F2 in codon space
y <- colSums(tuco) # Mass of codons
z <- as.data.frame(cbind(x, y))
rownames(z) <- colnames(tuco)
z <- z[order(z$x), ]
pt.cex <- sqrt(z$y)/40
xlab <- paste("F2:", signif(100*tuco.coa$eig[2]/sum(tuco.coa$eig), 3),
         "%")
icolor <- sapply(rownames(z), function(x) sum(s2c(x) %in% c("g", "c"))
         + 1)
color <- c(grey(1), grey(0.67), grey(0.33), grey(0))[icolor]
par(mar = c(5, 3, 1, 1))
dotchart(z$x, labels = rownames(z), pt.cex = pt.cex, bg = color,
         xlab = xlab)
```

A1.3.11. *Code for Figure 2.12*

```
mklab <- function(x, n){
  paste("F", n, ": ", signif(100*x$eig[n]/sum(x$eig), 3), "%", sep = "")
}
x <- tuco.coa$co[ , 3] # F2 in codon space
y <- colSums(tuco) # Mass of codons
z <- as.data.frame(cbind(x, y))
rownames(z) <- colnames(tuco)
z <- z[order(z$x), ]
pt.cex <- sqrt(z$y)/40
xlab <- mklab(tuco.coa, 3)
aaname <- sapply(rownames(z), function(x) translate(s2c(x)))
```

```
color <- character(nrow(z))
data("SEQINR.UTIL")
for(i in 1:length(aaname)){
  if(aaname[i] %in% SEQINR.UTIL$AA.PROPERTY$Charged) color[i]
       <- col2alpha("white")
  else if(aaname[i] %in% SEQINR.UTIL$AA.PROPERTY$Non.polar) color[i]
       <- col2alpha("black")
  else color[i] <- col2alpha(grey(0.5))
}
par(mar = c(5, 3, 1, 1))
dotchart(z$x, labels = rownames(z), pt.cex = pt.cex, bg = color,
       xlab = xlab)
```

A1.3.12. *Code for Figure 2.13*

```
par(mar = c(4, 4, 1, 0) + 0.1)
mosaicplot(riblea, shade = TRUE, main = "")
```

A1.3.13. *Code for Figure 2.14*

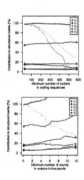

```
final.delrow <- final.delcol <- FALSE
xx <- sort(unique(rowSums(tuco)))
xx <- xx[1:(length(xx) - 85)] # At least 1/10 th CDS left
tuco.coa <- dudi.coa(tuco, scannf = FALSE, nf = 5)
(nrow(tuco) - 1)*(ncol(tuco) - 1)/sum(tuco) -> exptoti
pti2 <-100*tuco.coa$eig[1:5]/(sum(tuco.coa$eig) - exptoti)
```

```
par(mar = c(4, 4, 1, 1), mfrow = c(1, 2))
plot(rep(xx[1], 5), pti2, pch = 1:5, xlim = range(xx), ylim = c(0, 110),
    xlab = "Minimum number of codons\nin coding sequences", las = 1,
    ylab = "Contribution to structured inertia [%]")
if(final.delrow){
  res <- matrix(NA, nrow = length(xx), ncol = 7)
  for(i in 1:length(xx)){
    tuco2 <- tuco[rowSums(tuco) > xx[i], ]
    imp2 <- imp[rowSums(tuco) > xx[i]]
    tuco2.coa <- dudi.coa(tuco2, scannf = FALSE, nf = 5)
    (nrow(tuco2) - 1)*(ncol(tuco2) - 1)/sum(tuco2) -> exptoti
    res[i, 1:5] <- 100*tuco2.coa$eig[1:5]/(sum(tuco2.coa$eig) - exptoti)
    res[i, 6] <-100*sum(tuco2)/sum(tuco)
    res[i, 7] <-100*sum(tuco2[imp2, ])/sum(tuco[imp, ])
  }
  save(res, file = "local/delrow.rda")
} else {
  load("local/delrow.rda")
}
for(j in 1:5) points(xx, res[, j], type = "l")
points(xx, rowSums(res[, 1:5]), type = "l", lwd = 2)
points(rep(xx[length(xx)], 5), res[length(xx), 1:5], pch = 1:5)
points(xx, res[, 6], type = "l", col = "red", lty = 2)
points(xx, res[, 7], type = "l", col = "darkblue", lty = 4)
legend <- sapply(paste("lambda[", 1:5, "]", sep = ""),
        function(x) as.expression(parse(text = x)))
legend("topright", inset = 0.02, pch = 1:5, legend = legend, cex = 0.75,
  bg = grey(0.95))
# Right panel:
xx <- sort(unique(colSums(tuco)))/1000 # in kb
xx <- xx[1:(length(xx) - 6)] # At least 1/10 th codon left
tuco.coa <- dudi.coa(tuco, scannf = FALSE, nf = 5)
(nrow(tuco) - 1)*(ncol(tuco) - 1)/sum(tuco) -> exptoti
pti2 <-100*tuco.coa$eig[1:5]/(sum(tuco.coa$eig) - exptoti)
plot(rep(xx[1], 5), pti2, pch = 1:5, xlim = range(xx), ylim = c(0, 110),
    xlab = "Minimum number of counts\nin codons in thousands", las = 1,
    ylab = "Contribution to structured inertia [%]")
if(final.delcol){
  res <- matrix(NA, nrow = length(xx), ncol = 6)
  for(i in 1:length(xx)){
    tuco2 <- tuco[ , (colSums(tuco)/1000) > xx[i]]
    tuco2.coa <- dudi.coa(tuco2, scannf = FALSE, nf = 5)
    (nrow(tuco2) - 1)*(ncol(tuco2) - 1)/sum(tuco2) -> exptoti
    res[i, 1:5] <- 100*tuco2.coa$eig[1:5]/(sum(tuco2.coa$eig) - exptoti)
    res[i, 6] <- 100*sum(tuco2)/sum(tuco)
  }
  save(res, file = "local/delcol.rda")
} else {
  load("local/delcol.rda")
}
for(j in 1:5) points(xx, res[, j], type = "l", pch = j)
points(xx, rowSums(res[, 1:5]), type = "l", lwd = 2)
points(rep(xx[length(xx)], 5), res[length(xx), 1:5], pch = 1:5)
points(xx, res[, 6], type = "l", col = "red", lty = 2)
legend("topright", inset = 0.02, pch = 1:5, legend = legend, cex = 0.75,
  bg = grey(0.95))
```

A1.4. Chapter 3

A1.4.1. *Code for Figure 3.1*

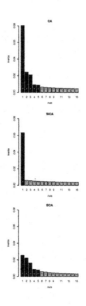

```
myscreeplot <- function(x, y, npcs = 15, nf = x$nf,
                    p.pch = 19, p.col = "red", p.cex = 0.75, ...){
  screeplot(x, npcs = npcs, ...) ; res.barplot <- barplot(x$eig,
          plot = FALSE)
  imax <- min(npcs, x$rank)
  points(res.barplot[1:imax, 1], y[1:imax], pch = p.pch, col = p.col,
          cex = p.cex)
}
par(mfrow = c(1, 3), mar = c(5, 4, 3, 0))
load("local/screeplot2.rda")
maxi <- ttuco.coa$eig[1]
myscreeplot(ttuco.coa, apply(sim, 1, mean)[1:15], ylim = c(0, maxi),
        main = "CA")
if(final.ttuco.eig){
  nreplicateswb <- 1000
  set.seed(1) # for reproducibility
  rs <- rowSums(ttuco) ; cs <- colSums(ttuco)
  fw <- function(){
    rtable <- r2dtable(1, rs, cs)[[1]]
    ttuco.coa <- dudi.coa(rtable, scan = FALSE, nf = 5)
    eig <- wca(ttuco.coa, facaa, scan = FALSE, nf = 1)$eig
    length(eig) <- 43
    return(eig)
  }
  simw <- replicate(nreplicateswb, fw())
  simw[which(is.na(simw), arr.ind = TRUE)] <- 0
  fb <- function(){
```

```
  rtable <- r2dtable(1, rs, cs)[[1]]
  ttuco.coa <- dudi.coa(rtable, scan = FALSE, nf = 5)
  eig <- bca(ttuco.coa, facaa, scan = FALSE, nf = 5)$eig
  length(eig) <- 20
  return(eig)
}
simb <- replicate(nreplicateswb, fb())
simb[which(is.na(simb), arr.ind = TRUE)] <- 0
save(nreplicateswb, simw, simb, file = "local/ttuco-eig.rda")
} else {
  load("local/ttuco-eig.rda")
}
myscreeplot(ttuco.wca, apply(simw, 1, mean)[1:15], ylim = c(0, maxi),
        main = "WCA")
myscreeplot(ttuco.bca, apply(simb, 1, mean)[1:15], ylim = c(0, maxi),
        main = "BCA")
```

A1.4.2. *Code for Figure 3.2*

```
mklab <- function(x, n){
  paste("F", n, ": ", signif(100*x$eig[n]/sum(x$eig), 3), "%", sep = "")
}
x <- ttuco.wca$co[, 1] ; xlab <- mklab(ttuco.wca, 1)
par(mar = c(4, 4, 1, 1) + 0.1, mfrow = c(1, 2))
hist(x, breaks = 30, las = 1, col = "lightblue", main = "", xlab = xlab,
  ylab = "Coding sequence count")
boxplot(x~leading, las = 1, varwidth = TRUE, horizontal = TRUE,
  names = c("lagging", "leading"), col = rev(collealag), xlab = xlab,
  pars = list(boxwex = 0.4, staplewex = 0.5, outwex = 0.5))
```

A1.4.3. *Code for Figure 3.3*

```
mklab <- function(x, n){
  paste("F", n, ": ", signif(100*x$eig[n]/sum(x$eig), 3), "%", sep = "")
}
```

```
x <- ttuco.wca$li[ , 1] # F1 in codon space
y <- colSums(tuco) # Mass of codons
z <- as.data.frame(cbind(x, y))
rownames(z) <- colnames(tuco)
z <- z[order(z$x), ]
pt.cex <- sqrt(z$y)/40
xlab <- mklab(ttuco.wca, 1)
color <- ifelse(substr(rownames(z), 3, 3) %in% c("g", "t"),
        col2alpha(grey(0.1)), "transparent")
par(mar = c(5, 3, 1, 1))
dotchart(z$x, labels = rownames(z), pt.cex = pt.cex, bg = color,
        xlab = xlab)
```

A1.4.4. *Code for Figure 3.4*

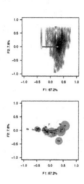

```
mklab <- function(x, n){
  paste("F", n, ": ", signif(100*x$eig[n]/sum(x$eig), 3), "%", sep = "")
}
baby <- rbind(tuco, rep(floor(sum(tuco)/64), 64)) # Adding a uniform
        codon usage CDS
babyaa <- t(apply(baby, 1, tapply, facaa, sum))
babyaa.coa <- dudi.coa(babyaa, scannf = FALSE)
xlab <- mklab(babyaa.coa, 1) ; ylab <- mklab(babyaa.coa, 2)
x <- babyaa.coa$li[, 1] ; xlim <- c(-1, 1)
y <- babyaa.coa$li[, 2] ; ylim <- xlim
par(mfrow = c(1, 2), mar = c(4, 4, 1, 1), lend = "butt")
smoothScatter(x, y, las = 1, xlab = xlab, ylab = ylab, xlim = xlim,
        ylim = ylim,
  colramp = colorRampPalette(c("white", "black")), nrpoints = 0)
myarrow <- function(){
  arrows(-0.3, 0, 0.3, 0, lwd = 3, angle = 20, length = 0.1)
  arrows(-0.29, 0, 0.3, 0, lwd = 1, angle = 20, length = 0.1,
        col = "white")
  text(0, 0.2, expression(F[0]), adj = c(0.45, 0.45))
  text(0, 0.2, expression(F[0]), adj = c(0.5, 0.5), col = "white")
}
myarrow()
# Right panel
x <- babyaa.coa$co[, 1]
```

```
y <- babyaa.coa$co[, 2]
cex <- sqrt(colSums(t(apply(tuco, 1, tapply, facaa, sum))))/25
bg <- rep("transparent", ncol(babyaa))
bg[colnames(babyaa) %in% c("Ile","Phe", "Lys", "Tyr", "Asn", "Leu")]
        <- col2alpha("red")
bg[colnames(babyaa) %in% c("Gly", "Pro", "Ala", "Arg")]
        <- col2alpha("blue")
plot(x, y, las = 1, xlab = xlab, ylab = ylab, cex = cex, xlim = xlim,
        ylim = ylim,
     pch = 21, col = grey(0.9), bg = bg)
myarrow()
text(x, y, colnames(babyaa), cex = 0.5)
```

A1.4.5. *Code for Figure 3.5*

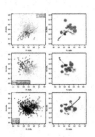

```
mklab <- function(x, n){
  paste("F", n, ": ", signif(100*x$eig[n]/sum(x$eig), 3), "%", sep = "")
}
x <- ttuco.bca$co[, 1] ; y <- ttuco.bca$co[, 2]
xlab <- mklab(ttuco.bca, 1) ; ylab <- mklab(ttuco.bca, 2)
isrib <- rownames(tuco) %in% ribnames
par(mar = c(4, 4, 1, 0) + 0.1, mfrow = c(3, 2))
# Panel (1, 1)
lim <- c(-0.6, 0.6)
plot(x, y, bg = "transparent", pch = 21, col = col2alpha(grey(0.8)),
        xlab = xlab,
   ylab = ylab, main = "", las = 1, xlim = lim, ylim = lim)
points(x[isrib], y[isrib], pch = pchrib, bg = colrib)
legend("topright", inset = 0.02, legend = c("ribosomal"),
   pt.bg = colrib, pch = pchrib, bg = grey(0.95))
arrows(0, 0, -0.4, -0.4, length = 0.1, angle = 10)
text(-0.4, -0.4, "G1", pos = 1)
# Panel (1, 2)
x <- ttuco.bca$li[, 1] ; y <- ttuco.bca$li[, 2]
xlab <- mklab(ttuco.bca, 1) ; ylab <- mklab(ttuco.bca, 2)
aanames <- rownames(ttuco.bca$tab)
cex <- sqrt(500*ttuco.bca$lw)
bg <- rep("transparent", length(aanames))
bg[aanames %in% c("Ile","Phe", "Lys", "Tyr", "Asn", "Leu")]
        <- col2alpha("red")
bg[aanames %in% c("Gly", "Pro", "Ala", "Arg")] <- col2alpha("blue")
plot(x, y, pch = 21, xlab = xlab, ylab = ylab, main = "", cex = cex,
   las = 1, xlim = lim, ylim = lim, bg = bg, col = col2alpha("grey"))
text(x, y, aanames, xpd = NA)
```

```
arrows(0, 0, -0.4, -0.4, length = 0.1, angle = 10)
text(-0.4, -0.4, "G1", pos = 1)
# Panel (2, 1)
x <- ttuco.bca$co[, 1] ; y <- ttuco.bca$co[, 3]
xlab <- mklab(ttuco.bca, 1) ; ylab <- mklab(ttuco.bca, 3)
data(EXP) ; kd <- (tuco/rowSums(tuco)) %*% EXP$KD ; imp <- kd > 0.5
plot(x, y, col = col2alpha(grey(0.8)), xlab = xlab, ylab = ylab,
        main = "",
   las = 1, xlim = lim, ylim = lim)
points(x[imp], y[imp], pch = pchimp, bg = colimp)
legend("bottomleft", inset = 0.02, legend = c("integral membrane
        protein"),
   pt.bg = colimp, pch = pchimp, bg = grey(0.95))
arrows(0, 0, -0.4, 0.4, length = 0.1, angle = 10)
text(-0.4, 0.4, "G2", pos = 3)
# Panel (2, 2)
x <- ttuco.bca$li[, 1] ; y <- ttuco.bca$li[, 3]
xlab <- mklab(ttuco.bca, 1) ; ylab <- mklab(ttuco.bca, 3)
bg <- rep(col2alpha(grey(0.5)), length = length(aanames))
data("SEQINR.UTIL")
bg[a(aanames) %in% SEQINR.UTIL$AA.PROPERTY$Charged]
        <- col2alpha("white")
bg[a(aanames) %in% SEQINR.UTIL$AA.PROPERTY$Non.polar]
        <- col2alpha("black")
plot(x, y, pch = 21, xlab = xlab, ylab = ylab, main = "", cex = cex,
   las = 1, asp = 1, xlim = lim, ylim = lim, bg = bg,
        col = col2alpha("grey"))
text(x, y, aanames, xpd = NA)
arrows(0, 0, -0.4, 0.4, length = 0.1, angle = 10)
text(-0.4, 0.4, "G2", pos = 3)
# Panel (3, 1)
x <- ttuco.bca$co[, 1] ; y <- ttuco.bca$co[, 3]
xlab <- mklab(ttuco.bca, 1) ; ylab <- mklab(ttuco.bca, 3)
bg <- ifelse(leading, collea, collag)
pch <- ifelse(leading, pchlea, pchlag)
par(mar = c(4, 4, 1, 0) + 0.1)
plot(x, y, pch = pch, bg = bg, xlab = xlab, ylab = ylab, main = "",
        las = 1,
   xlim = lim, ylim = lim)
legend("bottomleft", inset = 0.02, legend = c("leading", "lagging"),
   pt.bg = c(collea, collag), pch = c(pchlea, pchlag), bg = grey(0.95))
arrows(0, 0, 0.4, 0.4, length = 0.1, angle = 10)
text(0.4, 0.4, "G3", pos = 3)
# Panel (3, 2)
x <- ttuco.bca$li[, 1] ; y <- ttuco.bca$li[, 3]
xlab <- mklab(ttuco.bca, 1) ; ylab <- mklab(ttuco.bca, 3)
bg <- rep("transparent", length(aanames))
bg[aanames %in% c("Phe","Leu", "Val", "Gly", "Cys", "Trp")]
        <- col2alpha(collealag[1])
bg[aanames %in% c("Pro", "His", "Gln", "Thr", "Asn", "Lys")]
        <- col2alpha(collealag[2])
plot(x, y, pch = 21, xlab = xlab, ylab = ylab, main = "", cex = cex,
   las = 1, xlim = lim, ylim = lim, bg = bg, col = col2alpha("grey"))
text(x, y, aanames, xpd = NA)
arrows(0, 0, 0.4, 0.4, length = 0.1, angle = 10)
text(0.4, 0.4, "G3", pos = 3)
```

A1.5. Chapter 4

A1.5.1. *Code for Figure 4.1*

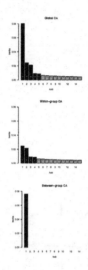

```
myscreeplot <- function(x, y, npcs = 15, nf = x$nf,
                        p.pch = 19, p.col = "red", p.cex = 0.75, ...){
  screeplot(x, npcs = npcs, ...) ; res.barplot <- barplot(x$eig,
           plot = FALSE)
  imax <- min(npcs, x$rank)
  points(res.barplot[1:imax, 1], y[1:imax], pch = p.pch, col = p.col,
         cex = p.cex)
}
par(mfrow = c(1, 3), mar = c(5, 4, 4, 1) + 0.1)
load("local/screeplot2.rda")
ylim <- c(0, tuco.coa$eig[1])
myscreeplot(tuco.coa, rowMeans(sim), main = "Global CA", ylim = ylim,
           las = 1)
if(final.ica.eig1){
  nreplicateswbg <- 1000
  set.seed(1) # for reproducibility
  rs <- rowSums(tuco) ; cs <- colSums(tuco)
  fwg <- function(){
    rtable <- r2dtable(1, rs, cs)[[1]]
    tuco.coa <- dudi.coa(rtable, scan = FALSE, nf = 5)
    eig <- wca(tuco.coa, as.factor(leading), scan = FALSE, nf = 4)$eig
    length(eig) <- min(dim(tuco)) - 1 # 63
    return(eig)
  }
  simwg <- replicate(nreplicateswbg, fwg())
  simwg[which(is.na(simwg), arr.ind = TRUE)] <- 0
  fbg <- function(){
    rtable <- r2dtable(1, rs, cs)[[1]]
    tuco.coa <- dudi.coa(rtable, scan = FALSE, nf = 5)
```

```
    eig <- bca(tuco.coa, as.factor(leading), scan = FALSE, nf = 1)$eig
    length(eig) <- length(levels(as.factor(leading))) - 1 # 1
    return(eig)
  }
  simbg <- replicate(nreplicateswbg, fbg())
  simbg[which(is.na(simbg), arr.ind = TRUE)] <- 0
  save(nreplicateswbg, simwg, simbg, file = "local/ica-eig1.rda")
} else {
  load("local/ica-eig1.rda")
}
# Second panel
myscreeplot(tuco.wca, rowMeans(simwg), main = "Within-group CA",
        ylim = ylim, las = 1)
# Third panel
myscreeplot(tuco.bca, mean(simbg), main = "Between-group CA",
        ylim = ylim, las = 1)
```

A1.5.2. *Code for Figure 4.2*

```
myscreeplot <- function(x, y, npcs = 15, nf = x$nf,
                    p.pch = 19, p.col = "red", p.cex = 0.75, ...){
  screeplot(x, npcs = npcs, ...) ; res.barplot <- barplot(x$eig,
        plot = FALSE)
  imax <- min(npcs, x$rank)
  points(res.barplot[1:imax, 1], y[1:imax], pch = p.pch, col = p.col,
        cex = p.cex)
}
par(mfrow = c(3, 3), mar = c(2, 2, 4, 0.5))
# Panel (1, 1)
load("local/screeplot2.rda")
ylim <- c(0, ttuco.coa$eig[1])
myscreeplot(ttuco.coa, y = rowMeans(sim), main = "Global codon usage",
        ylim = ylim)
# Panel (1, 2)
load("local/ttuco-eig.rda")
myscreeplot(ttuco.wca, rowMeans(simw), ylim = ylim, main = "Synonymous")
# Panel (1, 3)
myscreeplot(ttuco.bca, rowMeans(simb), ylim = ylim,
        main = "Non-synonymous")
# Panel (2, 1)
load("local/ica-eig1.rda")
myscreeplot(tuco.wca, rowMeans(simwg), main = "Within-group",
        ylim = ylim)
if(final.ica.eig2){
  nreplicateswbgwba <- 1000
  set.seed(1) # for reproducibility
  rs <- rowSums(ttuco) ; cs <- colSums(ttuco)
```

```
fwgwa <- function(){
  rtable <- r2dtable(1, rs, cs)[[1]]
  ttuco.coa <- dudi.coa(rtable, scan = FALSE, nf = 5)
  ttuco.wca <- wca(ttuco.coa, facaa, scan = FALSE, nf = 1)
  eig <- wca(t(ttuco.wca), facg, scan = F, nf = 0)$eig
  length(eig) <- 43
  return(eig)
}
simwgwa <- replicate(nreplicateswbgwba, fwgwa())
simwgwa[which(is.na(simwgwa), arr.ind = TRUE)] <- 0
fbgwa <- function(){
  rtable <- r2dtable(1, rs, cs)[[1]]
  ttuco.coa <- dudi.coa(rtable, scan = FALSE, nf = 5)
  ttuco.wca <- wca(ttuco.coa, facaa, scan = FALSE, nf = 1)
  eig <- bca(t(ttuco.wca), facg, scan = F, nf = 0)$eig
  length(eig) <- 1
  return(eig)
}
simbgwa <- replicate(nreplicateswbgwba, fbgwa())
simbgwa[which(is.na(simbgwa), arr.ind = TRUE)] <- 0
fwgba <- function(){
  rtable <- r2dtable(1, rs, cs)[[1]]
  ttuco.coa <- dudi.coa(rtable, scan = FALSE, nf = 5)
  ttuco.bca <- wca(ttuco.coa, facaa, scan = FALSE, nf = 1)
  eig <- wca(t(ttuco.bca), facg, scan = F, nf = 0)$eig
  length(eig) <- 20
  return(eig)
}
simwgba <- replicate(nreplicateswbgwba, fwgba())
simwgba[which(is.na(simwgba), arr.ind = TRUE)] <- 0
fbgba <- function(){
  rtable <- r2dtable(1, rs, cs)[[1]]
  ttuco.coa <- dudi.coa(rtable, scan = FALSE, nf = 5)
  ttuco.bca <- wca(ttuco.coa, facaa, scan = FALSE, nf = 1)
  eig <- bca(t(ttuco.bca), facg, scan = F, nf = 0)$eig
  length(eig) <- 1
  return(eig)
}
simbgba <- replicate(nreplicateswbgwba, fbgba())
simbgba[which(is.na(simbgba), arr.ind = TRUE)] <- 0

save(nreplicateswbgwba, simwgwa, simbgwa, simwgba, simbgba,
       file = "local/ica-eig2.rda")
} else {
  load("local/ica-eig2.rda")
}
# Panel (2, 2)
myscreeplot(ttuco.wca.wca, rowMeans(simwgwa),
       main = "Within-group\nSynonymous",
  ylim = ylim, col = "grey")
# Panel (2, 3)
myscreeplot(ttuco.bca.wca, rowMeans(simwgba),
  main = "Within-group\nNon-synonymous", ylim = ylim)
# Panel (3, 1)
myscreeplot(tuco.bca, mean(simbg), main = "Between-group", ylim = ylim)
# Panel (3, 2)
myscreeplot(ttuco.wca.bca, mean(simbgwa),
       main = "Between-group\nSynonymous",
  ylim = ylim)
# Panel (3, 3)
```

```
myscreeplot(ttuco.bca.bca, mean(simbgba),
    main = "Between-group\nNon-synonymous", ylim = ylim)
```

A1.5.3. *Code for Figure 4.3*

```
myscreeplot <- function(x, y, npcs = 15, nf = x$nf,
                        p.pch = 19, p.col = "red", p.cex = 0.75, ...){
  screeplot(x, npcs = npcs, ...) ; res.barplot <- barplot(x$eig,
        plot = FALSE)
  imax <- min(npcs, x$rank)
  points(res.barplot[1:imax, 1], y[1:imax], pch = p.pch, col = p.col,
        cex = p.cex)
}
par(mfrow = c(3, 3), mar = c(2, 2, 4, 0.5))
# Panel (1, 1)
load("local/screeplot2.rda")
ylim <- c(0, ttuco.coa$eig[1])
myscreeplot(ttuco.coa, y = rowMeans(sim), main = "Global codon usage",
        ylim = ylim)
# Panel (1, 2)
load("local/ttuco-eig.rda")
myscreeplot(ttuco.wca, rowMeans(simw), ylim = ylim, main = "Synonymous")
# Panel (1, 3)
myscreeplot(ttuco.bca, rowMeans(simb), ylim = ylim,
        main = "Non-synonymous")
# Panel (2, 1)
load("local/ica-eig1.rda")
myscreeplot(tuco.wca, rowMeans(simwg), main = "Within-group",
        ylim = ylim)
if(final.ica.eig2){
  nreplicateswbgwba <- 1000
  set.seed(1) # for reproducibility
  rs <- rowSums(ttuco) ; cs <- colSums(ttuco)
  fwgwa <- function(){
    rtable <- r2dtable(1, rs, cs)[[1]]
    ttuco.coa <- dudi.coa(rtable, scan = FALSE, nf = 5)
    ttuco.wca <- wca(ttuco.coa, facaa, scan = FALSE, nf = 1)
    eig <- wca(t(ttuco.wca), facg, scan = F, nf = 0)$eig
    length(eig) <- 43
    return(eig)
  }
  simwgwa <- replicate(nreplicateswbgwba, fwgwa())
  simwgwa[which(is.na(simwgwa), arr.ind = TRUE)] <- 0
  fbgwa <- function(){
    rtable <- r2dtable(1, rs, cs)[[1]]
    ttuco.coa <- dudi.coa(rtable, scan = FALSE, nf = 5)
    ttuco.wca <- wca(ttuco.coa, facaa, scan = FALSE, nf = 1)
```

```
      eig <- bca(t(ttuco.wca), facg, scan = F, nf = 0)$eig
      length(eig) <- 1
      return(eig)
    }
    simbgwa <- replicate(nreplicateswbgwba, fbgwa())
    simbgwa[which(is.na(simbgwa), arr.ind = TRUE)] <- 0
    fwgba <- function(){
      rtable <- r2dtable(1, rs, cs)[[1]]
      ttuco.coa <- dudi.coa(rtable, scan = FALSE, nf = 5)
      ttuco.bca <- wca(ttuco.coa, facaa, scan = FALSE, nf = 1)
      eig <- wca(t(ttuco.bca), facg, scan = F, nf = 0)$eig
      length(eig) <- 20
      return(eig)
    }
    simwgba <- replicate(nreplicateswbgwba, fwgba())
    simwgba[which(is.na(simwgba), arr.ind = TRUE)] <- 0
    fbgba <- function(){
      rtable <- r2dtable(1, rs, cs)[[1]]
      ttuco.coa <- dudi.coa(rtable, scan = FALSE, nf = 5)
      ttuco.bca <- wca(ttuco.coa, facaa, scan = FALSE, nf = 1)
      eig <- bca(t(ttuco.bca), facg, scan = F, nf = 0)$eig
      length(eig) <- 1
      return(eig)
    }
    simbgba <- replicate(nreplicateswbgwba, fbgba())
    simbgba[which(is.na(simbgba), arr.ind = TRUE)] <- 0

    save(nreplicateswbgwba, simwgwa, simbgwa, simwgba, simbgba,
         file = "local/ica-eig2.rda")
} else {
  load("local/ica-eig2.rda")
}
# Panel (2, 2)
myscreeplot(ttuco.wca.wca, rowMeans(simwgwa),
       main = "Within-group\nSynonymous",
  ylim = ylim, col = "grey")
# Panel (2, 3)
myscreeplot(ttuco.bca.wca, rowMeans(simwgba),
  main = "Within-group\nNon-synonymous", ylim = ylim)
# Panel (3, 1)
myscreeplot(tuco.bca, mean(simbg), main = "Between-group", ylim = ylim)
# Panel (3, 2)
myscreeplot(ttuco.wca.bca, mean(simbgwa),
       main = "Between-group\nSynonymous",
  ylim = ylim)
# Panel (3, 3)
myscreeplot(ttuco.bca.bca, mean(simbgba),
  main = "Between-group\nNon-synonymous", ylim = ylim)
```

A1.5.4. *Code for Figure 4.4*

```
mklab <- function(x, n){
  paste("F", n, ": ", signif(100*x$eig[n]/sum(x$eig), 3), "%", sep = "")
}
x <- ttuco.wca.bca$co[ , 1] # F1 in codon space
y <- colSums(tuco) # Mass of codons
z <- as.data.frame(cbind(x, y))
rownames(z) <- colnames(tuco)
z <- z[order(z$x), ]
pt.cex <- sqrt(z$y)/40
xlab <- mklab(ttuco.wca.bca, 1)
color <- ifelse(substr(rownames(z), 3, 3) %in% c("g", "t"),
        col2alpha(grey(0.1)), "transparent")
par(mar = c(5, 3, 1, 1))
dotchart(z$x, labels = rownames(z), pt.cex = pt.cex, bg = color,
        xlab = xlab)
```

A1.5.5. *Code for Figure 4.5*

```
mklab <- function(x, n){
  paste("F", n, ": ", signif(100*x$eig[n]/sum(x$eig), 3), "%", sep = "")
}
x <- ttuco.bca.bca$co[ , 1] # F1 in codon space
y <- ttuco.bca.bca$cw # Mass of aa
z <- as.data.frame(cbind(x, y))
rownames(z) <- rownames(ttuco.bca.bca$co)
z <- z[order(z$x), ]
pt.cex <- sqrt(200*z$y)
xlab <- mklab(ttuco.bca.bca, 1)
par(mar = c(5, 3, 1, 1))
bg <- rep(col2alpha("grey"), 21)
names(bg) <- rownames(z)
```

```
bg[names(bg) %in% c("Phe", "Val", "Gly", "Cys", "Trp") ]
      <- col2alpha("white")
bg[names(bg) %in% c("Pro", "His", "Gln", "Asn", "Lys", "Thr") ]
      <- col2alpha("black")
dotchart(z$x, labels = rownames(z), pt.cex = pt.cex, bg = bg,
      xlab = xlab)
```

A1.5.6. *Code for Figure 4.6*

```
mklab <- function(x, n){
  paste("F", n, ": ", signif(100*x$eig[n]/sum(x$eig), 3), "%", sep = "")
}
x <- ttuco.bca.wca$li[, 1] ; y <- ttuco.bca.wca$li[, 2]
xlab <- mklab(ttuco.bca.wca, 1) ; ylab <- mklab(ttuco.bca.wca, 2)
par(mar = c(4, 4, 1, 1))
plot(x, y, main = "", xlab = xlab, ylab = ylab, pch = 21,
      col = grey(0.7), las = 1)
data(EXP) ; kd <- (tuco/rowSums(tuco)) %*% EXP$KD ; imp <- kd > 0.45
points(x[imp], y[imp], bg = colimp, pch = pchimp)
rib <- rownames(tuco) %in% ribnames
points(x[rib], y[rib], bg = colrib, pch = pchrib)
legend("bottomright", inset = 0.02, c("Ribosomal proteins",
  "Integral membrane proteins"), pch = c(pchrib, pchimp),
  pt.bg = c(colrib, colimp), bg = grey(0.95))
```

A1.5.7. *Code for Figure 4.7*

```
mklab <- function(x, n){
  paste("F", n, ": ", signif(100*x$eig[n]/sum(x$eig), 3), "%", sep = "")
}
bgaagc <- function(aa){
  bg <- rep(col2alpha(grey(0.5)), length(aa))
  bg[aa == "Ile"] <- col2alpha(grey(1))
```

```
  bg[aa %in% c("Phe", "Lys", "Tyr", "Asn")] <- col2alpha(grey(0.95))
  bg[aa == "Leu"] <- col2alpha(grey(0.7))
  bg[aa == "Met"] <- col2alpha(grey(0.6))
  bg[aa == "Trp"] <- col2alpha(grey(0.4))
  bg[aa == "Arg"] <- col2alpha(grey(0.05))
  bg[aa %in% c("Gly", "Pro", "Ala")] <- col2alpha(grey(0))
  return(bg)
}
x <- ttuco.bca.wca$co[ , 1] # F1 in aa space
y <- ttuco.bca.wca$cw # Mass of aa
z <- as.data.frame(cbind(x, y))
rownames(z) <- rownames(ttuco.bca.wca$co)
z <- z[order(z$x), ]
pt.cex <- sqrt(200*z$y)
xlab <- mklab(ttuco.bca.wca, 1)
par(mar = c(5, 3, 1, 1))
col <- rep("transparent", 21)
dotchart(z$x, labels = rownames(z), pt.cex = pt.cex, xlab = xlab,
  bg = bgaagc(rownames(z)))
```

A1.5.8. *Code for Figure 4.8*

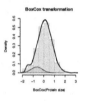

```
data(EXP) ; kd <- (tuco/rowSums(tuco)) %*% EXP$KD ; imp <- kd > 0.5
size <- rowSums(tuco) - 1
par(mfrow = c(1, 3))
myhist <- function(x, ...){
  dst <- density(x)
  hst <- hist(x, plot = FALSE)
```

```
hist(x, probability = TRUE, border = grey(0.8), col = grey(0.9),
   ylim = c(0, max(dst$y, hst$density)), ...)
lines(density(x), lwd = 2)
dst2 <- density(x[imp])
polygon(dst2$x, sum(imp)*dst$y/nrow(tuco), col = colimp)
}
myhist(size, main = "Original data", xlab = "Protein size [#aa]")
#
library(MASS)
res <- boxcox(size~1, lambda = seq(0.1, 0.5, length = 1000),
        plotit = FALSE)
lest <- res$x[which.max(res$y)]
fboxcox <- function(y, lambda){
  stopifnot(all(y > 0))
  y <- y/exp(mean(log(y)))
  if(isTRUE(all.equal(lambda, 0))){
    log(y)
  } else {
    (y^lambda - 1)/lambda
  }
}
myhist(fboxcox(size, lest), main = "BoxCox transformation",
  xlab = "BoxCox(Protein size)")
myhist(log10(size), main = "Logarithmic transformation",
  xlab = "Log10(Protein size)")
```

A1.5.9. *Code for Figure 4.9*

```
mklab <- function(x, n){
  paste("F", n, ": ", signif(100*x$eig[n]/sum(x$eig), 3), "%", sep = "")
}
fx <- 3 ; fy <- 4
dudi <- ttuco.bca.wca ; x <- dudi$li[, fx] ; y <- dudi$li[, fy]
xlab <- mklab(dudi, fx) ; ylab <- mklab(dudi, fy)
xlim <- range(dudi$li[, fx], dudi$co[, fx])
ylim <- range(dudi$li[, fy], dudi$co[, fy])
par(mar = c(4, 4, 1, 1))
smoothScatter(x, y, main = "", xlab = xlab, ylab = ylab,
  las = 1, xlim = xlim, ylim = ylim, asp = 1, nbin = 256,
  colramp = colorRampPalette(c("white", "black")), nrpoints = 0)
text(dudi$co[, fx], dudi$co[, fy], rownames(dudi$co), cex = 1.5,
      col = "firebrick")
sml <- rowSums(tuco) < 75
points(x[sml], y[sml], pch = 21, bg = "white")
```

Appendix 2

A2.1. Session information

```
sessionInfo()
```

```
R version 3.4.1 (2017-06-30)
Platform: x86_64-apple-darwin15.6.0 (64-bit)
Running under: macOS Sierra 10.12.6
Matrix products: default
BLAS: /Library/Frameworks/R.framework/Versions/3.4/Resources/lib/libRblas.0.dylib
LAPACK: /Library/Frameworks/R.framework/Versions/3.4/Resources/lib/libRlapack.dylib

locale:
[1] fr_FR.UTF-8/fr_FR.UTF-8/fr_FR.UTF-8/C/fr_FR.UTF-8/fr_FR.UTF-8

attached base packages:
[1] stats     graphics  grDevices utils     datasets  methods   base

other attached packages:
 [1] xtable_1.8-2    plot3D_1.1.1    vioplot_0.2     sm_2.2-5.4      gplots_3.0.1
 [6] MASS_7.3-47     dichromat_2.0-0 fortunes_1.5-4  rgl_0.98.1      seqinr_3.4-5
[11] ade4_1.7-8

loaded via a namespace (and not attached):
 [1] Rcpp_0.12.11    knitr_1.17      magrittr_1.5    misc3d_0.8-4
 [5] R6_2.2.2        caTools_1.17.1  tools_3.4.1     KernSmooth_2.23-15
 [9] htmltools_0.3.6 gtools_3.5.0    digest_0.6.12   shiny_1.0.5
[13] htmlwidgets_1.2 bitops_1.0-6    mime_0.5        gdata_2.18.0
[17] compiler_3.4.1  jsonlite_1.5    httpuv_1.3.5
```

References

[ADE 14] ADEOLU M., GUPTA R., "A phylogenomic and molecular marker based proposal for the division of the genus *Borrelia* into two genera: the emended genus *Borrelia* containing only the members of the relapsing fever *Borrelia*, and the genus *Borreliella* gen. nov. containing the members of the Lyme disease *Borrelia* (*Borrelia burgdorferi* sensu lato complex)", *Antonie van Leeuwenhoek*, vol. 105, no. 6, pp. 1049–1072, 2014.

[ADL 14] ADLER D., MURDOCH D., rgl: 3D visualization device system (OpenGL), R package version 0.93.996, 2014.

[AKA 02] AKASHI H., GOJOBORI T., "Metabolic efficiency and amino acid composition in the proteomes of *Escherichia coli* and *Bacillus subtilis*", *Proceedings of the National Academy of Sciences of the United States of America*, vol. 99, pp. 3695–3700, 2002.

[AND 90] ANDERSSON S., KURLAND C., "Codon preferences in free-living microorganisms", *Microbiological Reviews*, vol. 54, pp. 198–210, 1990.

[AND 98] ANDERSSON S., ZOMORODIPOUR A., ANDERSSON J., SICHERITZ-PONTEN T., ALSMARK U., PODOWSKI R., NASLUND A., ERIKSSON A., WINKLER H., KURLAND C., "The genome sequence of *Rickettsia prowazekii* and the origin of mitochondria", *Nature*, vol. 396, pp. 133–140, 1998.

[BAK 92] BAKER T., WICKNER S., "Genetics and enzymology of DNA replication in *Escherichia coli*", *Annual Review of Genetics*, vol. 26, pp. 447–477, 1992.

[BAR 93] BARBOUR A., "Linear DNA of Borrelia species and antigenic variation", *Trends in Microbiology*, vol. 1, no. 6, pp. 236–239, 1993.

[BAR 17] BARBOUR A., ADEOLU M., GUPTA R., "Division of the genus *Borrelia* into two genera (corresponding to Lyme disease and relapsing fever groups) reflects their genetic and phenotypic distinctiveness and will lead to a better understanding of these two groups of microbes (Margos *et al.* (2016) There is inadequate evidence to support the division of the genus *Borrelia*. *Int. J. Syst. Evol. Microbiol.* doi: 10.1099/ijsem.0.001717)", *International Journal of Systematic and Evolutionary Microbiology*, vol. 67, pp. 2058–2067, 2017.

[BÉC 05] BÉCUE M., PAGÈS J., PARDO C.-E., "Contingency table with a double partition on rows and colums. Visualization and comparison of the partial and global structures", in JANSSEN J., LENCA P. (eds), *Applied Stochastic Models and Data Analysis*, ENST Bretagne, Brest, pp. 355–364, 2005.

[BEL 58] BELOZERSKY A., SPIRIN A., "A correlation between the compositions of deoxyribonucleic and ribonucleic acids", *Nature*, vol. 182, pp. 111–112, 1958.

[BEN 73] BENZÉCRI J.-P., *L'analyse des données. II: L'analyse des correspondances*, Dunod, Paris, 1973.

[BEN 83] BENZÉCRI J.-P., "Analyse de l'inertie intra-classe par l'analyse d'un tableau des correspondances", *Les Cahiers de l'Analyse des Données*, vol. 8, pp. 351–358, 1983.

[BHA 16] BHAGWAT A., WEILONG HAO W., TOWNES J., LEE H., TANG H., FOSTER P., "Strand-biased cytosine deamination at the replication fork causes cytosine to thymine mutations in *Escherichia coli*", *Proceedings of the National Academy of Sciences of the United States of America*, vol. 113, pp. 2176–2181, 2016.

[BLA 97] BLATTNER F., PLUNKETT III G., BLOCH C., PERNA N., BURLAND V., RILLEY M., COLLADO-VIDES J., GLASNER J., RODE C., MAYHEW G., GREGOR J., DAVIS N., KIRKPATRICK H., GOEDEN M., ROSE D., MAU B., SHAO Y., "The complete genome sequence of *Escherichia coli* K-12", *Science*, vol. 277, no. 5331, pp. 1453–1462, 1997.

[BOX 64] Box G., Cox D., "An analysis of transformations", *Journal of the Royal Statistical Society, B*, vol. 26, pp. 211–252, 1964.

[BRE 88] Brewer B., "When polymerase collide: replication and the transcriptional organization of the *E. coli* chromosome", *Cell*, vol. 53, pp. 679–686, 1988.

[BRO 82] Brown G., Simpson M., "Novel features of animal mtDNA evolution as shown by sequences of two rat cytochrome oxidase subunit II genes", *Proceedings of the National Academy of Sciences of the United States of America*, vol. 79, pp. 3246–3250, 1982.

[CAS 00] Casjens S., Palmer N., van Vugt R., Huang W., Stevenson B., Rosa P., Lathigra R., Sutton G., Peterson J., Dodson R., Haft D., Hickey E., Gwinn M., White O., Fraser C., "A bacteria genome in flux: the twelve linear and nine circular extrachromosomal DNAs in an infectious isolate of Lyme disease spirochete *Borrelia burgdorferi*", *Molecular Microbiology*, vol. 35, pp. 490–516, 2000.

[CAT 66] Cattell R., "The scree test for the number of factors", *Multivariate Behavioral Research*, vol. 1, pp. 245–276, 1966.

[CAZ 88] Cazes P., Chessel D., Dolédec S., "L'analyse des correspondances internes d'un tableau partitionné: son usage en hydrobiologie", *Revue de Statistique Appliquée*, vol. 36, pp. 39–54, 1988.

[CEB 98] Cebrat S., Dudek M., "The effect of DNA phase structure on DNA walks", *The European Physical Journal B*, vol. 3, pp. 271–276, 1998.

[CEB 99] Cebrat S., Dudek M., Gierlik A., Kowalczuk M., Mackiewicz P., "Effect replication on the third base of codons", *Physica A*, vol. 265, pp. 78–84, 1999.

[CHA 05] Charif D., Thioulouse J., Lobry J., Perrire G., "Online synonymous codon usage analyses with the ade4 and seqinR packages", *Bioinformatics*, vol. 21, no. 4, pp. 545–7, 2005.

[CHA 07] CHARIF D., LOBRY J., "SeqinR 1.0-2: a contributed package to the R project for statistical computing devoted to biological sequences retrieval and analysis", in BASTOLLA U., PORTO M., ROMAN H.E., VENDRUSCOLO M. (eds), *Structural Approaches to Sequence Evolution: Molecules, Networks, Populations*, Biological and Medical Physics, Biomedical Engineering, Springer Verlag, New York, pp. 207–232, 2007.

[CHE 04] CHESSEL D., DUFOUR A.-B., THIOULOUSE J., "The ade4 package – I: one-table methods", *R News*, vol. 4, pp. 5–10, 2004.

[CLE 85] CLEVELAND W., *The Elements of Graphing Data*, Wadsworth, Monterey, CA, 1985.

[DRA 07] DRAY S., DUFOUR A., "The ade4 package: implementing the duality diagram for ecologists", *Journal of Statistical Software*, vol. 22, pp. 1–20, 2007.

[EME 98] EMERSON J. W., "Mosaic displays in S-PLUS: a general implementation and a case study", *Statistical Computing and Graphics Newsletter (ASA)*, vol. 9, pp. 17–23, 1998.

[EME 10] EMERY L., Codon usage bias in Archaea, PhD Thesis, University of Edinburgh, 2010.

[ERM 01] ERMOLAEVA M., "Synonymous codon usage in bacteria", *Current Issues in Molecular Biology*, vol. 3, pp. 91–97, 2001.

[ESC 78] ESCOFIER B., "Analyse factorielle et distances répondant au principe d'équivalence distributionnelle", *Revue de Statistique Appliquée*, vol. 26, pp. 29–37, 1978.

[FRA 97] FRASER C., CASJENS S., HUANG W., SUTTON G., CLAYTON R., LATHIGRA R., WHITE O., KETCHUM K., DODSON R., HICKEY E., GWINN M., DOUGHERTY B., TOMB J., FLEISCHMANN R., RICHARDSON D., PETERSON J., KERLAVAGE A., QUACKENBUSH J., SALZBERG S., HANSON M., VAN VUGT R., PALMER N., ADAMS M., GOCAYNE J., WEIDMAN J., UTTERBACK T., WATTHEY L., MCDONALD L., ARTIACH P., BOWMAN C., GARLAND S., FUJI C., COTTON M., HORST K., ROBERTS K., HATCH B., SMITH H., VENTER J., "Genomic sequence of a Lyme disease spirochaete, *Borrelia burgdorferi*", *Nature*, vol. 390, pp. 580–586, 1997.

[FRA 99] FRANK A., LOBRY J., "Asymmetric substitution patterns: a review of possible underlying mutational or selective mechanisms", *Gene*, vol. 238, pp. 65–77, 1999.

[FRE 90] FREDERICO L., KUNKEL T., SHAW B., "A sensitive genetic assay for the detection of cytosine deamination: determination of rate constants and the activation energy", *Biochemistry*, vol. 29, pp. 2532–2537, 1990.

[FRE 98] FREEMAN J., PLASTERER T., SMITH T., MOHR S., "Patterns of genome organization in bacteria", *Science*, vol. 279, pp. 1827–1827, 1998.

[FRI 94] FRIENDLY M., "Mosaic displays for multi-way contingency tables", *Journal of the American Statistical Association*, vol. 89, pp. 190–200, 1994.

[GAO 17] GAO N., LU G., LERCHER M., CHEN W.-H., "Selection for energy effciency drives strand-biased gene distribution in prokaryotes", *Nature Scientific Reports*, vol. 7, p. 10572, 2017.

[GAR 16] GARCÍA-MUSE T., AGUILERA A., "Transcription-replication conflicts: how they occur and how they are resolved", *Nature Review Molecular Cellular Biology*, vol. 17, pp. 553–563, 2016.

[GAU 82a] GAUTIER C., GOUY M., JACOBZONE M., GRANTHAM R., *Nucleic Acid Sequences Handbook*, Praeger Publishers, London, UK, vol. 1, 1982.

[GAU 82b] GAUTIER C., GOUY M., JACOBZONE M., GRANTHAM R., *Nucleic Acid Sequences Handbook*, Praeger Publishers, London, UK, vol. 2, 1982.

[GAU 87] GAUTIER C., Analyses statistiques et évolution des séquences d'acides nucléiques, PhD thesis, Université Claude Bernard - Lyon I, 1987.

[GAU 00] GAUTIER C., "Compositional bias in DNA", *Current Opinion in Genetics & Develoment*, vol. 10, pp. 656–661, 2000.

[GOU 82] GOUY M., GAUTIER C., "Codon usage in bacteria: correlation with gene expressivity", *Nucleic Acids Research*, vol. 10, pp. 7055–7073, 1982.

[GOU 84] GOUY M., MILLERET F., MUGNIER C., JACOBZONE M., GAUTIER C., "ACNUC: a nucleic acid sequence data base and analysis system", *Nucleic Acids Research*, vol. 12, pp. 121–127, 1984.

[GOU 85a] GOUY M., GAUTIER C., ATTIMONELLI M., LANAVE C., DI PAOLA G., "ACNUC – a portable retrieval system for nucleic acid sequence databases: logical and physical designs and usage", *Computer Applications in the Biosciences*, vol. 1, pp. 167–172, 1985.

[GOU 85b] GOUY M., GAUTIER C., MILLERET F., "System analysis and nucleic acid sequence banks", *Biochimie*, vol. 67, pp. 433–436, 1985.

[GOU 08] GOUY M., DELMOTTE S., "Remote access to ACNUC nucleotide and protein sequence databases at PBIL", *Biochimie*, vol. 90, pp. 555–562, 2008.

[GRA 80a] GRANTHAM R., GAUTIER C., GOUY M., "Codon frequencies in 119 individual genes confirm consistent choices of degenerate base according to genome type", *Nucleic Acids Research*, vol. 8, pp. 1892–1912, 1980.

[GRA 80b] GRANTHAM R., GAUTIER C., GOUY M., MERCIER R., PAV A., "Codon catalog usage and the genome hypothesis", *Nucleic Acids Research*, vol. 8, pp. r49–r62, 1980.

[GRE 84] GREENACRE M., *Theory and Applications of Correspondence Analysis*, Academic Press, London, 1984.

[GRI 98a] GRIGORIEV A., "Analyzing genomes with cumulative skew diagrams", *Nucleic Acids Research*, vol. 26, pp. 2286–2290, 1998.

[GRI 98b] GRIGORIEV A., FREEMAN J., PLASTERER T., SMITH T., MOHR S., "Genome arithmetic", *Science*, vol. 281, pp. 1923–1924, 1998.

[HAE 94] HAECKEL E., *Systematische Phylogenie. Entwurf eines natrlichen Systems der Organismen auf Grund ihrer Stammesgeschichte. Erster Theil, Systematische Phylogenie der Protisten und Pflanzen*, Georg Reimer, Berlin, 1894.

[HAM 16] HAMPERL S., CIMPRICH K., "Conflict resolution in the genome: how transcription and replication make it work", *Cell*, vol. 167, pp. 1455–1467, 2016.

[HAR 84] HARTIGAN J., KLEINER B., "A mosaic of television ratings", *The American Statistician*, vol. 38, pp. 32–35, 1984.

[HIN 98] HINTZE J. L., NELSON R. D., "Violin plots: a box plot-density trace synergis", *The American Statistician*, vol. 52, pp. 181–184, 1998.

[HOL 85] HOLMES-JUNCA S., Outils informatiques pour l'évaluation de la pertinence d'un résultat en analyse des données, PhD Thesis, Université des sciences et techniques du Languedoc, Académie de Montpellier, 1985.

[HOL 86] HOLM L., "Codon usage and gene expression", *Nucleic Acids Research*, vol. 27, pp. 244–247, 1986.

[HOL 08] HOLMES S., "Multivariate analysis: the French way", *Probability and Statistics: Essays in Honor of David A. Freedman*, vol. 2, pp. 219–233, IMS Lecture Notes – Monograph Series, 2008.

[HOW 16] HOWARD M., "A review of exploratory factor analysis decisions and overview of current practices: what we are doing and how can we improve?", *International Journal of Human-Computer Interaction*, vol. 32, pp. 51–62, 2016.

[IKE 81] IKEMURA T., "Correlation between the abundance of *Escherichia coli* transfer RNAs and the occurrence of the respective codons in its protein genes", *Journal of Molecular Biology*, vol. 146, pp. 1–21, 1981.

[JOE 18] JOESCH-COHEN L., ROBINSON M., JABBARI N., LAUSTED C., GLUSMAN G., "Novel metrics for quantifying bacterial genome composition skews", *BMC Genomics*, vol. 19, pp. 528–540, 2018.

[KAI 58] KAISER H. F., "The varimax criterion for analytic rotation in factor analysis", *Psychometrika*, vol. 23, pp. 187–200, 1958.

[KAN 99] KANAYA S., YAMADA Y., KUDO Y., IKEMURA T., "Studies of codon usage and tRNA genes of 18 unicellular organisms and quantification of *Bacillus subtilis* tRNA: gene expression level and species-specific diversity of codon usage based on multivariate analysis", *Gene*, vol. 238, pp. 143–155, 1999.

[KAR 98] KARLIN S., CAMPBELL A., MRÁZEK J., "Comparative DNA analysis across diverse genomes", *Annual Review of Genetics*, vol. 23, pp. 185–225, 1998.

[KAR 99] KARLIN S., "Bacterial DNA strand compositional asymmetry", *Trends in Microbiology*, vol. 7, pp. 305–308, 1999.

[KOW 01] KOWALCZUK M., MACKIEWICZ P., MACKIEWICZ D., NOWICKA A., DUDKIEWICZ M., DUDEK M., CEBRAT S., "DNA asymmetry and the replicational mutational pressure", *Journal of Applied Genetics*, vol. 42, pp. 553–577, 2001.

[KOW 02] KOWALCZUK M., Presja mutacyjna i selekcyjna w genomie *Borrelia burgdorferi*, PhD Thesis, Wroclaw University, 2002.

[KUH 62] KUHN T., *The Structure of Scientific Revolutions*, University of Chicago Press, Chicago, 1962.

[KYT 82] KYTE J., DOOLITTLE R., "A simple method for displaying the hydropathic character of a protein", *Journal of Molecular Biology*, vol. 157, pp. 105–132, 1982.

[LAF 99] LAFAY B., LLOYD A., MCLEAN M., DEVINE K., SHARP P., WOLFE K., "Proteome composition and codon usage in spirochaetes: species-specific and DNA strand-specific mutational biases", *Nucleic Acids Research*, vol. 27, pp. 1642–1649, 1999.

[LEI 02] LEISCH F., "Sweave: dynamic generation of statistical reports using literate data analysis", in HÄRDLE W., RÖNZ B. (eds), *Compstat 2002 – Proceedings in Computational Statistics*, Physica Verlag, Heidelberg, pp. 575–580, 2002.

[LIN 74] LINDAHL T., NYBERG B., "Heat-induced deamination of cytosine residues in deoxy-ribonucleic acid", *Biochemistry*, vol. 13, pp. 3405–3410, 1974.

[LIU 95] LIU B., ALBERTS M., "Head-on collision between a DNA replication apparatus and RNA polymerase transcription complex", *Science*, vol. 267, pp. 1131–1137, 1995.

[LOB 94] LOBRY J., GAUTIER C., "Hydrophobicity, expressivity and aromaticity are the major trends of amino-acid usage in 999 *Escherichia coli* chromosome-encoded genes", *Nucleic Acids Research*, vol. 22, pp. 3174–3180, 1994.

[LOB 96a] LOBRY J., "Asymmetric substitution patterns in the two DNA strands of bacteria", *Molecular Biology and Evolution*, vol. 13, pp. 660–665, 1996.

[LOB 96b] LOBRY J., "Origin of replication of *Mycoplasma genitalium*", *Science*, vol. 272, no. 5262, pp. 745–746, 1996.

[LOB 96c] LOBRY J., "A simple vectorial representation of DNA sequences for the detection of replication origins in bacteria", *Biochimie*, vol. 78, pp. 323–326, 1996.

[LOB 97] LOBRY J., "Influence of genomic G+C content on average amino-acid composition of proteins from 59 bacterial species", *Gene*, vol. 205, pp. 309–316, 1997.

[LOB 00] LOBRY J., The black hole of symmetric molecular evolution. Habilitation thesis, PhD Thesis, Université Claude Bernard - Lyon 1, 2000.

[LOB 02] LOBRY J., SUEOKA N., "Asymmetric directional mutation pressures in bacteria", *Genome Biology*, vol. 3, no. 10, pp. research0058.1–research0058.14, 2002.

[LOB 03] LOBRY J., CHESSEL D., "Internal correspondence analysis of codon and amino-acid usage in thermophilic bacteria", *Journal of Applied Genetics*, vol. 44, pp. 235–261, 2003.

[LOB 06] LOBRY J., NECŞULEA A., "Synonymous codon usage and its potential link with optimal growth temperature in prokaryotes", *Gene*, vol. 385, pp. 128–136, 2006.

[LOP 99] LOPEZ P., PHILIPPE H., MYLLYKALLIO H., FORTERRE P., "Identification of putative chromosomal origins of replication in Archaea", *Molecular Microbiology*, vol. 32, pp. 883–891, 1999.

[LOP 01] LOPEZ P., PHILIPPE H., "Composition strand asymmetries in prokaryotic genomes: mutational bias and biased gene orientation", *Comptes Rendus de L'Académie des Sciences de Paris, Sciences de la vie*, vol. 324, no. 3, pp. 201–208, 2001.

[LUC 14] LUCAS A., amap: another multidimensional analysis package, R package version 0.8-14, 2014.

[LUM 13] LUMLEY T., dichromat: color schemes for dichromats, R package version 2.0-0, 2013.

[LUO 18] LUO H., QUAN C.-L., PENG C., GAO F., "Recent development of Ori-Finder system and DoriC database for microbial replication origins", *Briefings in Bioinformatics*, vol. 2018, pp. 1–11, 2018.

[LÊ 08] LÊ S., JOSSE J., HUSSON F., "FactoMineR: a package for multivariate analysis", *Journal of Statistical Software*, vol. 25, no. 1, pp. 1–18, 2008.

[MA 18] MA X.-X., MA P., CHANG Q.-Y., LIU Z.-B., ZHANG D., ZHOU X.-K., MA Z.-R., CAO X., "Adaptation of *Borrelia burgdorferi* to its natural hosts by synonymous codon and amino acid usage", *Journal of Basic Microbiology*, vol. 2018, pp. 1–11, 2018.

[MAC 99a] MACKIEWICZ P., GIERLIK A., KOWALCZUK M., DUDEK M., CEBRAT S., "Asymmetry of nucleotide composition of prokaryotic chromosomes", *Journal of Applied Genetics*, vol. 40, pp. 1–14, 1999.

[MAC 99b] MACKIEWICZ P., GIERLIK A., KOWALCZUK M., DUDEK M., CEBRAT S., "How does replication-associated mutational pressure influence amino acid composition of proteins?", *Genome Research*, vol. 9, pp. 409–416, 1999.

[MAC 99c] MACKIEWICZ P., GIERLIK A., KOWALCZUK M., SZCZEPANIK D., DUDEK M., CEBRAT S., "Mechanisms generating long-range correlation in nucleotide composition of the *Borrelia burgdorferi* genome", *Physica A*, vol. 273, pp. 103–115, 1999.

[MAO 12] MAO X., ZHANG H., YIN Y., XU Y., "The percentage of bacterial genes on leading versus lagging strands is influenced by multiple balancing forces", *Nucleic Acids Res*, vol. 40, pp. 8210–8218, 2012.

[MAR 92] MARIANS K., "Prokaryotic DNA replication", *Annual Review of Biochemistry*, vol. 61, pp. 673–719, 1992.

[MCI 98] MCINERNEY J., "Replication and transcriptional selection on codon usage in *Borrelia burgdorferi*", *Proceedings of the National Academy of Sciences of the United States of America*, vol. 95, pp. 10698–10703, 1998.

[MCL 98] MCLEAN M., WOLFE K., DEVINE K., "Base composition skews, replication orientation, and gene orientation in 12 prokaryote genomes", *Journal of Molecular Evolution*, vol. 47, pp. 691–696, 1998.

[MER 12] MERRIKH H., ZHANG Y., GROSSMAN A., WANG J., "Replication-transcription conflicts in bacteria", *Nature Review Microbiology*, vol. 10, pp. 449–458, 2012.

[MRÁ 98] MRÁZEK J., KARLIN S., "Strand compositional asymmetry in bacterial and large viral genomes", *Proceedings of the National Academy of Sciences of the United States of America*, vol. 95, pp. 3720–3725, 1998.

[NEN 07] NENADIC O., GREENACRE M., "Correspondence analysis in R, with two- and three-dimensional graphics: the ca package", *Journal of Statistical Software*, vol. 20, no. 3, pp. 1–13, 2007.

[NIS 80] NISHISATO S., *Analysis of Caregorical Data: Dual Scaling and Its Applications*, University of Toronto Press, London, UK, 1980.

[NOS 83] NOSSAL N.G., "Prokaryotic DNA replication systems", *Annual Review of Biochemistry*, vol. 53, pp. 581–615, 1983.

[OCH 02] OCHMAN H., "Distinguishing the ORFs from the ELFs: short bacterial genes and the annotation of genomes", *Trends in Genetics*, vol. 18, pp. 335–337, 2002.

[OLI 96] OLIVER J., MARIN A., "A relationship between GC content and coding-sequence length", *Journal of Molecular Evolution*, vol. 43, pp. 216–223, 1996.

[PAT 81] PATEFIELD W., "Algorithm AS159. An efficient method of generating r x c tables with given row and column totals", *Applied Statistics*, vol. 30, pp. 91–97, 1981.

[PEA 00] PEARSON K., "On the criterion that a given system of deviations from the probable in the case of correlated system of variables is such that it can be reasonably supposed to have arisen from random sampling", *The London, Edinburgh, and Dublin Philosophical Magazine and Journal of Science*, vol. 50, pp. 157–175, 1900.

[PED 99] PEDEN J., Analysis of codon usage, PhD Thesis, University of Nottingham, 1999.

[PEN 09] PENEL S., ARIGON A., DUFAYARD J., SERTIER A., DAUBIN V., DURET L., GOUY M., PERRIÉRE G., "Databases of homologous gene families for comparative genomics", *BMC Bioinformatics*, vol. 10, p. S3, 2009.

[PER 98] PERRIÈRE G., LOBRY J., "Asymmetrical coding sequence repartition and codon adaptation index values between leading and lagging strands in seven bacterial species", in KOLCHANOV N., SOLOVYEV V. (eds), *First International Conference on Bioinformatics of Genome Regulation and Structure*, Novosibirsk, Russia, vol. 2, pp. 254–255, 1998.

[PER 00] PERRIÈRE G., BESSIÈRES P., LABEDAN B., "EMGLib: the enhanced microbial genomes library (update 2000)", *Nucleic Acids Research*, vol. 28, pp. 68–71, 2000.

[PER 02] PERRIÈRE G., THIOULOUSE J., "Use and misuse of correspondence analysis in codon usage studies", *Nucleic Acids Research*, vol. 30, pp. 4548–4555, 2002.

[PER 05] PERES-NETO P., JACKSON D., SOMERS K., "How many principal components? Stopping rules for determining the number of non-trivial axes revisited", *Computational Statistics and Data Analysis*, vol. 49, pp. 974–997, 2005.

[PIC 99] PICARDEAU M., LOBRY J., HINNEBUSCH B., "Physical mapping of an origin of bidirectional replication at the centre of the *Borrelia burgdorferi* linear chromosome", *Molecular Microbiology*, vol. 32, pp. 437–445, 1999.

[RCO 13] R CORE TEAM, R: a language and environment for statistical computing, Software, R Foundation for Statistical Computing, Vienna, Austria. Available at http://www.R-project.org, 2013.

[ROC 99a] ROCHA E., DANCHIN A., VIARI A., "Bacterial DNA strand compositional asymmetry: response", *Trends in Microbiology*, vol. 7, pp. 308–308, 1999.

[ROC 99b] ROCHA E., DANCHIN A., VIARI A., "Universal replication biases in bacteria", *Molecular Microbiology*, vol. 32, pp. 11–16, 1999.

[SAL 98] SALZBERG S., SALZBERG A., KERLAVAGE A., TOMB J.-F., "Skewed oligomers and origins of replication", *Gene*, vol. 217, pp. 57–67, 1998.

[SÉM 06] SÉMON M., LOBRY J., DURET L., "No evidence for tissue-specific adaptation of synonymous codon usage in humans", *Molecular Biology and Evolution*, vol. 23, no. 3, pp. 523–529, 2006.

[SER 08] SERNOVA N., GELFAND M., "Identification of replication origins in prokaryotic genomes", *Briefings in Bioinformatics*, vol. 2008, pp. 1–16, 2008.

[SHA 86] SHARP P., TUOHY T., MOSURSKI K., "Codon usage in yeast: cluster analysis clearly differentiates highly and lowly expressed genes", *Nucleic Acids Research*, vol. 14, pp. 5125–5143, 1986.

[SHA 87] SHARP P., LI W.-H., "The codon adaptation index – a measure of directional synonymous codon usage bias, and its potential applications", *Nucleic Acids Research*, vol. 15, pp. 1281–1295, 1987.

[SHE 94] SHEN J.-C., RIDEOUT III. W.M., JONES P.A., "The rate of hydrolytic deamination of 5-methylcytosine in double-stranded DNA", *Nucleic Acids Research*, vol. 22, no. 6, pp. 972–976, 1994.

[SHP 89] SHPAER E.G., "Amino acid composition is correlated with protein abundance in *Escherichia coli*: can this be due to optimization of translational efficiency?", *Protein Sequences and Data Analysis*, vol. 2, no. 2, pp. 107–110, 1989.

[SUE 61] SUEOKA N., "Correlation between base composition of deoxyribonucleic acid and amino acid composition of protein", *Proceedings of the National Academy of Sciences of the United States of America*, vol. 47, pp. 1141–1149, 1961.

[SUE 62] SUEOKA N., "On the genetic basis of variation and heterogeneity of DNA base composition", *Proceedings of the National Academy of Sciences of the United States of America*, vol. 48, pp. 582–592, 1962.

[SUE 88] SUEOKA N., "Directional mutation pressure and neutral molecular evolution", *Proceedings of the National Academy of Sciences of the United States of America*, vol. 85, pp. 2653 –2657, 1988.

[SUE 95] SUEOKA N., "Intrastrand parity rules of DNA base composition and usages biases of synonymous codons", *Journal of Molecular Evolution*, vol. 40, no. 3, pp. 318–325, 1995.

[SUE 96] SUEOKA N., "Erratum: Intrastrand parity rules of DNA base composition and usages biases of synonymous codons", *Journal of Molecular Evolution*, vol. 42, p. 323, 1996.

[SUN 87] SUN MICROSYSTEMS, XDR: External Data Representation Standard, RFC 1014, Report, Network Working Group, 1987.

[SUZ 08] SUZUKI H., BROWN C.J., FORNEY L.J., TOP E.M., "Comparison of correspondence analysis methods for synonymous codon usage in bacteria", *DNA Research*, vol. 15, no. 6, pp. 357–365, 2008.

[TEK 16] TEKAIA F., "Genome data exploration using correspondence analysis", *Bioinformatics and Biology Insights*, vol. 10, pp. 59–72, 2016.

[TIL 00] TILLIER E., COLLINS R., "The contribution of replication orientation, gene direction, and signal sequences to base composition asymmetries in bacterial genomes", *Journal of Molecular Evolution*, vol. 50, pp. 249–257, 2000.

[VAN 05] VAN DE VELDEN M., "Rotation in correspondence analysis", *Journal of Classification*, vol. 22, no. 2, pp. 251–271, 2005.

[VEN 02] VENABLES W., RIPLEY B., *Modern Applied Statistics with S*, 4th edition, Springer, New York, 2002.

[WAR 87] WARTENBERG D., FERSON S., ROHLF F., "Putting things in order: a critique of detrended correspondence analysis", *The American Naturalist*, vol. 129, pp. 434–448, 1987.

[WAR 16] WARNES G., BOLKER B., BONEBAKKER L., GENTLEMAN R., LIAW W., LUMLEY T., MAECHLER M., MAGNUSSON A., MOELLER S., SCHWARTZ M., VENABLES B., gplots: Various R Programming Tools for Plotting Data, R package version 3.0.1, 2016.

[XIA 12] XIA X., "DNA Replication and strand asymmetry in prokaryotic and mitochondrial genomes", *Current Genomics*, vol. 13, pp. 16–27, 2012.

[ZEI 90] ZEIGLER D., DEAN D., "Orientation of genes in the *Bacillus subtilis* chromosome", *Genetics*, vol. 125, pp. 703–708, 1990.

[ZHE 15] ZHENG W.-X., LUO C.-S., DEN Y.-Y., GUO F., "Essentiality drives the orientation bias of bacterial genes in a continuous manner", *Nature Scientific Reports*, vol. 5, p. 16431, 2015.

[ZHO 14] ZHOU H.-Q., NING L.-W., ZHANG H.-X., GUO F.-B., "Analysis of the Relationship between genomic GC content and patterns of base usage, codon usage and amino acid usage in prokaryotes: similar GC content adopts similar compositional frequencies regardless of the phylogenetic lineages", *PLoS ONE*, vol. 9, p. e107319, 2014.

Index

A, B

ACNUC, 25
between-block correspondence
  analysis (BCA), 59–62, 66,
  67, 69, 71–75

C, D

codon adaptation index (CAI),
  26, 30, 48, 49, 65, 66
correspondence analysis (CA),
  1 , 10, 12, 16–21, 34, 36, 38,
  40–43, 46–50, 52–54, 59, 60,
  62, 65, 69, 71, 74, 80
deanimation, 44
degeneracy, 13, 69, 80

E, G

EMGLib, 25, 26, 31, 33, 94
expressivity, 30, 48, 51, 52,
  65–67, 80, 82, 87
genetic code, 60, 82
GRAVY score, 45
greedy, 55, 77, 82

I, K

inertia, 12, 21, 35–37, 54, 55,
  57, 69, 74, 75, 77
internal correspondence
  analysis (ICA), 71, 76, 77
kernel density, 82

L, M

lagging, 26, 41–44, 54, 63, 67,
  71, 74–76, 80
leading, 26, 30, 41–44, 51–54,
  63, 67, 71–75, 80
membrane, 45, 46, 48, 51, 55,
  57, 82, 84, 85, 87, 88
metabolic cost, 80
metric, 1, 2, 8, 10, 20, 21
missing factor, 38–40, 66, 67
mutation pressure, 44, 65, 66,
  87

N, O

non-synonymous, 22, 24, 59,
  62, 66, 74, 75, 79, 82
outliers, 83

P, R

principle component analysis
   (PCA), 1, 20
ribosomal, 30–33, 46, 48,
   51–54, 68, 82

S, W

scree plot, 12–15, 35, 36, 40,
   44, 53, 55, 56, 62, 72, 74–77
structured, 38, 54, 55, 57, 62,
   72, 74–77
synonymous, 22, 24, 27, 34, 44,
   59–63, 65, 66, 74, 75, 78, 79,
   82, 87
within-block correspondence
   analysis (WCA), 59, 60, 62,
   63, 64, 65, 71, 72, 74, 75

Printed in the United States
By Bookmasters